THE COMPLETE

THE COMPLETE

ICE AGE

How Climate Change Shaped the World

Edited by Brian Fagan

With 192 illustrations, 160 in color

Thames & Hudson

half-title Tip of a reconstructed Neanderthal spear.
title page Nineteenth-century engraving of a mammoth skeleton, the 'Missouri Leviathan'.
pages 4–5 Upper Palaeolithic painting from Lascaux cave in southwestern France, depicting animals including aurochs, horses and reindeer.

First published in 2009 in hardcover in the United States of America by
Thames & Hudson Inc., 500 Fifth Avenue,
New York, New York 10110

thamesandhudsonusa.com

Library of Congress Catalog Card Number 2008911980

ISBN 978-0-500-05161-0

Printed and bound in China by Toppan Printing

CONTENTS

Introducing the Ice Age

ICE AGE: the words conjure up images of huge glaciers, snow-clad mountains and ponderous mammoths grazing on windy steppes. There is something compelling about an ancient, deep-frozen world where fur-clad hunters stalked long-extinct beasts and subzero winters lasted for nine months a year. The Victorians discovered the Ice Age at a time when climatology was in its infancy and geological science confined, for the most part, to the study of stratified layers of the earth. They lived in an era when winters were often colder than they are today, which made it easier to think of long millennia of unremitting cold. *The Complete Ice Age*, written during a time of what may prove to be prolonged warming, paints a very different portrait of the last geological epoch, known to geologists as the Pleistocene.

We visit a world where there were, indeed, numerous long glacial periods, but one where fluctuations between colder and warmer times never ceased. For more than three quarters of the past 800,000 years global climate has been in transition between cold and warm conditions. In fact, climate changes were the backdrop to human history from the days when the first toolmaking hominins flourished in a drying Africa over 2 million years ago, towards the beginning of the Ice Age. Much of this book is concerned with the ways in which our remote, and more recent, forebears adapted to these extraordinary fluctuations. From their lives we may gain an understanding that provides us with important clues as we work out how to survive in our own warming world.

The Complete Ice Age reviews state-of-the-art knowledge. This is a journey transformed for us by new generations of multidisciplinary science that involve not only climatology and geology, but refined archaeological excavation and surveys. Much of the research described in these pages comes not from the field, but from painstaking laboratory work – dissecting the laminated sediments of deep-sea cores, laboriously counting pollen grains, or comparing data from annual growth rings in ancient trees. Ancient DNA provides significant clues on human population movements, sometimes in response to Ice Age climate change, while generations of palaeoanthropologists have studied the minutiae of fossil hominin remains. It is only now, 170 years after a young Swiss scientist called Louis Agassiz announced the discovery of the Ice Age, that we are beginning to understand its full complexity.

The portrait of the Ice Age presented here comes from the pens of four scholars, all of them actively involved in different aspects of Ice Age research. Brian Fagan describes how the Ice Age was discovered and how scientists from the Victorian geologist and palaeontologist William Buckland onwards built on the pioneer work of Louis Agassiz to confirm its existence. In a second chapter, he reviews the chronological investigations of the remarkable self-taught 19th-century Scottish scientist James Croll and the prescient calculations of Milutin Milankovitch, who spent a lifetime creating a mathematical theory to explain ancient and modern climates. Milankovitch's theories – still highly relevant today – dominated Ice Age research for a generation in the mid-20th century until two new dating methods,

The Aletsch glacier, some 10 km (6 miles) long, up to 2 km (1.2 miles) wide and 900 m (2,900 ft) thick, is the largest in the Alps today. But at the last Ice Age peak, roughly 20,000 years ago, much of northernmost Europe, North America and southern South America was covered by huge, contiguous glaciers (or ice sheets) 3 to 4 km (1.8 to 2.5 miles) thick.

Much of what we know today about ancient climate comes from ice cores drawn from the world's remaining ice sheets. Here curator Geoffrey Hargreaves inspects core samples from the Greenland ice sheet, which are stored in a freezer at –36°C (–33°F). The cores will be examined for evidence of Ice Age climate change, and also for signs of global warming caused by rising carbon dioxide levels.

radiocarbon and potassium-argon, provided the first relatively accurate timescale for the Ice Age. By the early 1960s, it was apparent that the epoch had begun at least 1 million years ago. Since then, new research has pushed the beginning of the Pleistocene back to about 2.5 million years, by which time gradual cooling was affecting many parts of the world, including the cradle of humankind, Africa.

Climatologist Mark Maslin now takes up the story, with a discussion of what we know today about the beginnings and causes of the Ice Age. He places the Ice Age in a broader geological context and provides us with a chronological baseline for later events. He then describes the dramatic climatic rollercoaster of the Pleistocene, especially during the past 800,000 years, when at least nine glacial periods were separated by shorter periods of rapid warming.

During the cold snaps, huge ice sheets mantled most of what is now Canada and Scandinavia. Northern winters lasted for nine months, with temperatures below –20°C (–4°F) for weeks on end. Summer temperatures were at least 10°C (18°F) lower than present ones. Global sea levels plunged to at least 90 m (295 ft) below those of today – a simply mind-boggling amount by comparison with the less than 1 m (3 ft) rise predicted for the next 100 years, devastating though that will be for modern populations. A land bridge joined Siberia and Alaska; large continental shelves extended from the Pacific and Atlantic coasts of North America and from Southeast Asia; Britain was not an island. The shorter, warm interglacials witnessed dramatic global warming; as ice sheets shrank, temperate vegetation spread northwards and sea levels climbed rapidly. Above all, this chapter emphasizes the volatility of late Ice Age climate, now revealed by new research into the past 100,000 years. Once thought to be unremittingly cold, it turns out to have been a period of rapid climatic shifts, where at times the climate in Europe was almost as warm as today.

The seesawing Ice Age climate created extraordinary challenges and opportunities for the humans and animals that inhabited the Pleistocene world. Archaeologist John Hoffecker tells the human story. When the Ice Age began, somewhat ape-like hominins were the only humans on earth. They walked upright, made and used tools, foraged in small bands and were well adapted to the more open country of cooler, drier times. At first, humans were purely African mammals, but around 1.8 to 1.7 million years ago people suddenly appeared in Eurasia and Asia, living as far north as 40 degrees. By this time, we were no longer purely tropical animals.

Hoffecker also documents another movement out of Africa, perhaps some 750,000 years ago, which brought the ancestors of the Neanderthals to Europe. They carried with them distinctive handaxe technology and also knowledge of fire,

Three palaeolithic stone tools manufactured by Neanderthal hunters, found in the Moravia region, Czech Republic. Advanced technologies meant that Neanderthals were efficient hunters of mammoth and woolly rhino, enabling them to subsist off a diet high in protein and fat – essential for survival on the cold plains of eastern Europe.

so crucial for survival in colder northern latitudes. He describes how pre-modern humans in Asia, and presumably in Europe as well, swung with the climatic punches, moving far southwards into the subtropics during cold periods, northwards as conditions grew warmer. It was not until the last glacial period that humans began to adapt to much colder conditions. The European Neanderthals were the first people to settle the vast plains of eastern Europe during the last interglacial

some 125,000 years ago. One would expect them to have moved southwards when cold conditions returned some 100,000 years ago, but despite this a few bands appear to have lived year-round in relatively sheltered river valleys and as far east as the Altai Mountains of Central Asia. These Neanderthals were able to survive not so much because of their anatomical adaptations to extreme cold – although they had some, such as a very stocky build – but because of more effective hunting

At roughly 30,000 years old, this female skull (and associated finds) found in a cave in Gibraltar represents the latest known population of Neanderthals. Modern humans had been coexisting with the Neanderthals in Europe for 15,000 years, but the exact nature of the biological and cultural relationship is not clear. Did we out-compete them, pushing them to the fringes of the continent, where they eventually died out? Or did a sudden cold snap lead to their demise? These questions remain unresolved.

technology that enabled them to subsist off a diet of protein- and fat-rich meat, especially larger animals like the mammoth and woolly rhinoceros.

Anatomically modern humans, *Homo sapiens*, evolved in tropical Africa between 200,000 and 150,000 years ago and spread rapidly out of their homeland after 100,000 years ago, and in earnest about 50,000 years before the present. They moved into a bewildering array of habitats and climatic zones, from extreme Ice Age environments in Eurasia and Siberia to tropical forests and islands off Southeast Asia. They had reached Europe by 42,000 years ago and Australia around the same time. Hoffecker describes how the first modern humans on the Eurasian steppe adapted to volatile climate change by developing new technological solutions such as snares, nets and light hunting weapons that enabled them to take prey out of the reach of their Neanderthal predecessors, who eventually became extinct. By 40,000 to 30,000 years ago, people were paying at least brief summer visits to locations above the Arctic Circle, extreme environments to which their innovations such as sewn clothing helped them adapt.

Palaeontologist Hannah O'Regan describes the exotic and formidable Ice Age bestiary – the sabre-toothed cat, the fearsome cave bear confronted by humans in dark caves, and the woolly mammoth that towered up to 4 m (13 ft) above the ground, an awesome adversary for Neanderthals and other humans who tried to hunt them. Then there were bison and wild horses, reindeer and mastodons, all animals that required expert stalking skills and effective weaponry to attack. O'Regan points out that the animals we encounter today are the impoverished remnants of a huge diversity of creatures that existed until about 11,000 years ago. She describes some of the great animal dispersals of the Ice Age that were possible because of the emergence of continental shelves and land bridges during glacial periods, and fluctuating rainfall levels that made deserts like the Sahara into barriers and then highways. Like their human counterparts, the Ice Age bestiary adapted constantly to climatic

The Ice Age was not simply a period of unremitting cold; during the interglacial periods when the ice sheets retreated warm-adapted animals moved much further north than we might expect. These remains of hippopotamus (*above*) and a rhino (*left*) were found in Britain – a long way from the African and Asian environments their modern-day ancestors now inhabit.

opposite Although not carnivorous, the cave bear, which stood over 1 m (4 ft) tall and lived throughout Europe and Western Asia, would have been a fearsome beast for Ice Age humans to encounter.

change; sometimes their distributions expanded, sometimes they shrank. The big question, however – and one of the great enigmas of the Ice Age – is why so many big-game animals died out in the late Pleistocene. Was it simply due to climate change or were human hunters the main culprits? Hannah O'Regan sifts the evidence and comes up with answers.

Australian Dennis Collaton with a mammoth tusk, New Siberian Islands, Arctic Russia. The permafrost in some parts of the globe, which has not thawed since the end of the last glacial period, has resulted in the near-perfect preservation of a number of Ice Age animals.

By the closing stages of the Ice Age some 15,000 years ago, as Brian Fagan shows, warming began, and tiny numbers of hunter-gatherers moved into extreme northeast Siberia, which they had visited only during summers in earlier times. At some point – the date is uncertain – handfuls of these people hunted their way across the Bering land bridge and colonized the Americas for the first time in one of the final chapters of the great diaspora of modern humanity. The last gasp of the Ice Age came with a millennium-long severe cold snap 13,000 years ago that plunged much of Europe and North America into near-glacial conditions once more. Paradoxically, it seems to have been this short, sharp shock that helped stimulate the greatest innovation in human history – the development of agriculture. In southwest Asia especially – the Levant and Euphrates Valley – where human populations had expanded and permanent settlements grown up in the more benign conditions before the cold snap, people now faced a shrinking of habitats where wild cereals on which they depended had flourished. Their answer over time was to innovate and start cultivating those same cereals and to domesticate herd animals such as sheep and goats. Human ingenuity knows no geographical boundaries, and similar processes occurred in different parts of the world where conditions were right, such as the Indus Valley, China's Yangtze Valley and somewhat later in the Americas. Within a few millennia, many thousands of people were living in cities under the rule of divine and powerful leaders, in the world's first literate civilizations.

Yet climate change did not cease with the melting of the glaciers, and the last 10,000 years have seen a number of climatic oscillations. The so-called Medieval Warm Period (AD 800–1250) caused prolonged droughts in the Americas while the Little Ice Age (AD 1300–1860) ushered in increased storminess and periods of intense cold. Today, however, as Mark Maslin shows in a powerful final chapter, it is the profligacy and very success of modern humans that are threatening our planet's future through global warming; but will our long-term future be hot or cold? We hope that *The Complete Ice Age* will provide not only a powerful story of how humans, animals and climate have interacted over hundreds of thousands of years, but also a better understanding of the long-term trends that govern life on earth and how such knowledge might help our species survive for thousands of years yet.

A polar bear amidst thawing northern ice floes. Global warming threatens the survival of many Arctic animals such as the polar bear, while rising sea levels menace low-lying coasts and islands.

1

Discovering the Ice Age

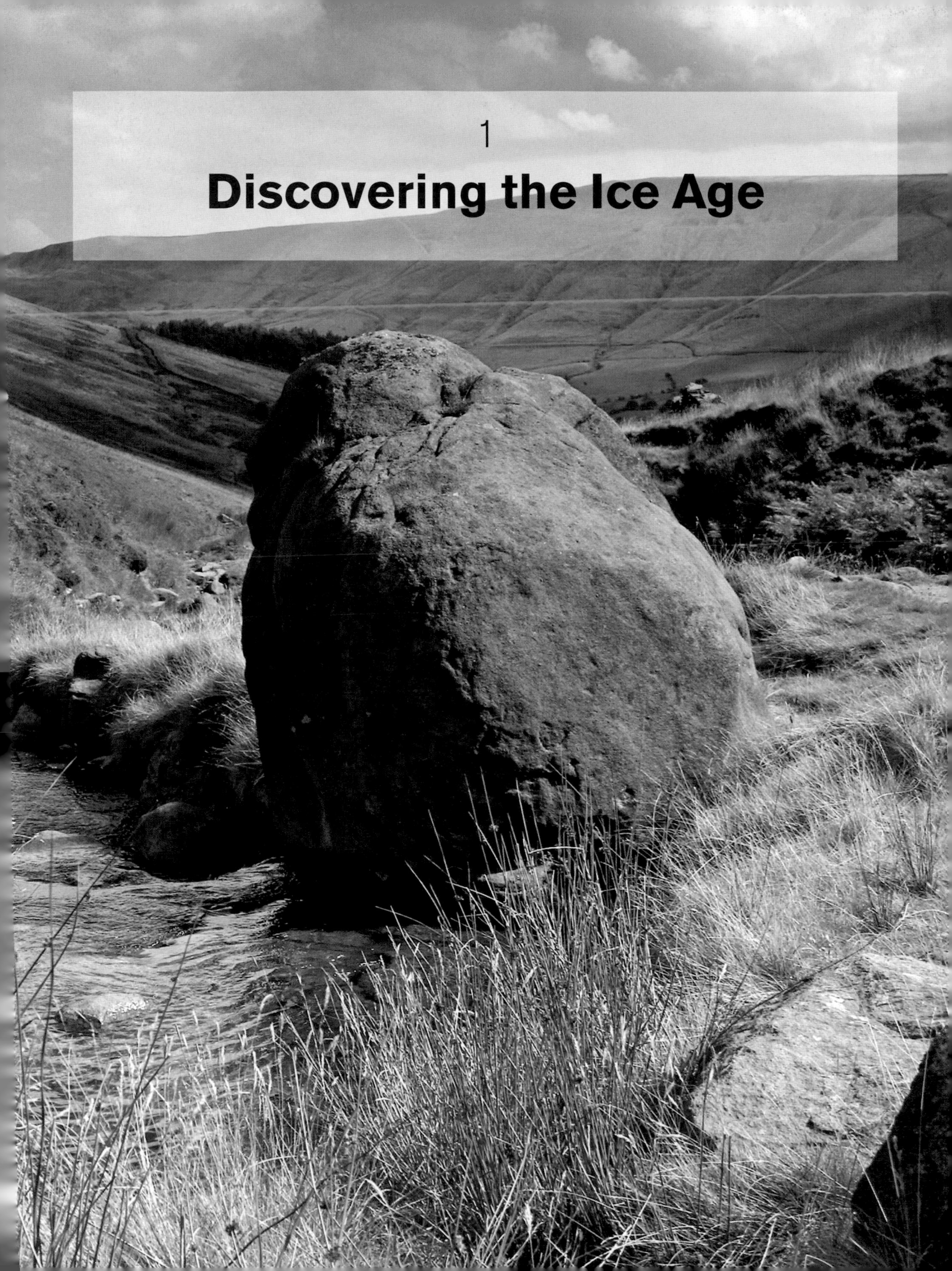

On 24 July 1837, the staid and highly respectable members of the Swiss Society of Natural Science gathered in Neuchâtel to hear an address from their young president, Louis Agassiz. They expected a lecture on the fossil fishes of Brazil, for Agassiz had acquired a stellar reputation for his research on this obscure subject. They were startled and outraged to instead receive a discourse on the grooved and much polished boulders of the nearby Jura Mountains. These erratic rocks, deposited far from their places of origin, had long been a geological mystery. In his lecture Agassiz boldly claimed to have solved the mystery, stating that great ice sheets had transported them, during an age of ice. What became known as the Discourse of Neuchâtel became the catalyst for a geological controversy over the existence of an Ice Age that raged for much of the 19th century.

previous pages An erratic boulder deposited by an ice sheet in the Edale Valley in England's Peak District. For years the forces that had moved such boulders so far from their source were a mystery, and many thought they had been transported by the Great Flood of the Bible.

above left Louis Agassiz (1807–1873), a pioneer of Ice Age science, at the Unteraar glacier, Switzerland, painted by Alfred Berthoud.

Controversy and observations

The leading scientists of the day rejected Agassiz's theory out of hand. The young scientist was a voluble, flamboyant personality, given to overstatement and bold ideas. However, many Swiss, who lived in intimate association with glaciers and their geological deposits, did in fact assume that a huge ice sheet had once covered the Alps, and amateur geologists had long reported signs of ancient glaciation from Scandinavia and the Alps. In 1793 the English geologist James Hutton, author of the groundbreaking book *Theory of the Earth*, visited the Jura and found clear evidence of primordial glacial activity, and in 1832 Reinhart Bernhardi, a German professor of natural science, argued that a polar ice cap had once extended as far south as central Germany.

A view of glacial erratics in the St Lawrence River at the Richelieu Rapids, drawn by Lieutenant Bowen in the spring of 1836. As in Europe, dramatic evidence of the Ice Age can be seen in many parts of North America.

Switzerland's Grindelwald glacier. In the 19th century this glacier stretched right into the valley. Thanks to prolonged warming, today's Alpine glaciers are much reduced since Louis Agassiz's day.

These theories, and many others, came independently, mainly from solitary field observations, but they made little headway in the face of prevailing scientific belief. The pervasive shackles of religious doctrines that argued for the literal historical truth of the scriptures held geology in a vice-like grip. *Genesis* 1 proclaimed that the Lord had created the earth and all living things in six days, and on the seventh he rested. The biblical chronology developed by the Irish clergyman James Ussher, Archbishop of Armagh, and others in the 17th century dated the Creation to 4004 BC, leaving only 6,000 years for all of geological time and human history. To challenge these teachings was heresy – and heresy was taken seriously in the early 19th century. The leading geologists of the 1830s, such as Charles Lyell of *Principles of Geology* fame, believed in the Universal Deluge, the last, cataclysmic biblical flood that had wiped out all of humanity and now-extinct animals – except for Noah and his ark. Under this rubric, the erratic boulders of the Alps and elsewhere had been transported by boulder-laden icebergs and ice rafts carried by the great inundation.

Agassiz had developed his ideas from the work of several observers, among them a mountaineer named Jean-Pierre Perraudin, who wrote of rocks scarred by the weight of moving glaciers. A highway engineer, Ignace Venetz, presented a paper to the Swiss Natural History Society in 1829, in which he used erratic boulders to argue that a huge ice sheet once covered the entire Alps. His ideas were ignored. Naturalist Jean de Charpentier, a salt mine manager, accumulated numerous observations and presented a glacial theory to the Society at Lucerne in 1834. Agassiz heard the paper and was unconvinced, until he spent a summer with Charpentier at Bex, where he looked at the evidence first hand. He visited glaciers with his host and Venetz and rapidly became a vigorous proponent of the glacial theory.

overleaf Louis Agassiz based his theories on extraordinarily detailed observations. He drew this annotated diagram of the Zermatt Glacier for his classic *Studies on Glaciers*, which was published in 1840. Opposite it is a view of the same glacier from the same publication without his annotations.

The ardent crusader

Whereas Charpentier was content to accumulate evidence, Agassiz was a passionate advocate, who soon outstripped the cautious seven years of observations of his colleagues. He developed a theory of an *Eiszeit*, an Ice Age, when a huge polar ice sheet enveloped all of Europe. The Discourse of Neuchâtel laid out his grandiose theory to a sceptical audience. Even a field trip to the Jura failed to convince his opponents. Nor did his monumental book *Studies on Glaciers*, which appeared in 1840. Alexander von Humboldt, doyen of natural scientists, in a letter urged his young colleague to return to his fishes, rather than persisting with his 'general considerations (a little icy besides) on the revolutions of the primitive world'.

Louis Agassiz was a man of vigorous imagination and eloquent prose. His ideas – scenarios of great ice sheets that crushed the animals and lush tropical world of earlier times – soon reached a wide audience. 'The silence of death followed ... sunrays rising over that frozen shore were met only by the whistling of northern winds and the rumbling of crevasses as they opened up across the surface of that huge ocean of ice.' Agassiz's Ice Age was as big a catastrophe as the Universal Deluge so beloved of many of his colleagues. His Ice Age theory might have been extravagant, but it did not run in the face of established religious dogma. One of those who listened carefully was the eccentric Oxford geologist William Buckland, who went into the field wearing his academic robes and a top hat. For all his eccentricity and devotion to the Universal Deluge, Buckland was a shrewd observer. He visited Agassiz in Switzerland in 1838 to see the glacial evidence at first hand. Initially Buckland was unconvinced, but a field trip with Agassiz to Scotland in 1840 changed his mind, for he could not otherwise explain how a flood

below Dean William Buckland (1784–1856) was a much beloved, devout geologist, seen here in a Thomas Sopwith cartoon fully equipped for a day among the glaciers. An expert field observer, Buckland was a believer in the literal historical truth of the scriptures, but accepted Agassiz's Ice Age theories.

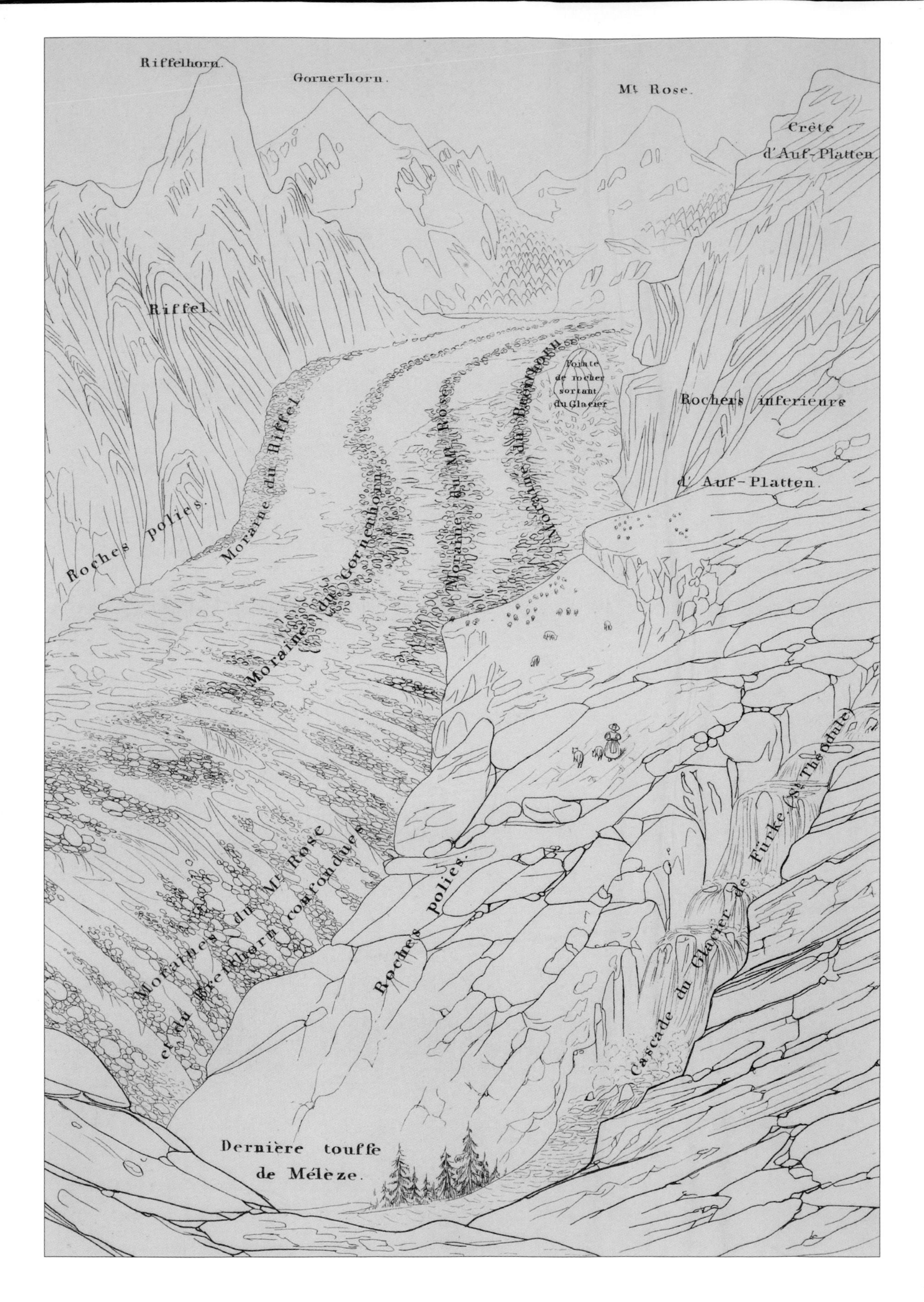
Riffelhorn.
Gornerhorn.
Mt Rose.
Crête d'Auf-Platten
Riffel.
Pointe de rocher sortant du Glacier
Rochers inferieurs d'Auf-Platten.
Roches polies.
Moraine du Riffel
Moraine du Gornerhorn
Moraine du Mt Rose
Moraine du Breithorn
Moraines du Mt Rose et du Breithorn confondues
Roches polies.
Cascade du Glacier de Furke (St Théodule)
Dernière touffe de Mélèze.

would have transported erratic boulders to altitudes of 1,524 m (5,000 ft) or more above sea level. He soon persuaded Charles Lyell of the existence of an Ice Age, by showing him, as he wrote in a letter to Agassiz in 1840, 'a beautiful cluster of moraines within two miles of his father's house'. In 1841 the geologist Edward Forbes wrote to Agassiz: 'You have made all the geologists glacier-mad here, and they are turning Great Britain into an ice house.'

Nevertheless, there was still powerful opposition to the idea of an Ice Age, not so much from religious orthodoxy, but from a profoundly conservative scientific establishment. One cannot entirely blame them, for there did appear to be ample geological evidence that the oceans had inundated much of the earth's continents on numerous occasions in the remote past. Agassiz's fossil fish were, after all, evidence for this. Many glacial deposits were devoid of fish fossils and marine shells, but some contained them, a conundrum that was not resolved until 1865, when Scottish geologist James Croll showed that they came from ice sheets that had moved over shallow water and scraped up fossilized sea shells. Furthermore, almost no geologists had first-hand experience of glaciers.

Louis Agassiz tended to exaggerate his discoveries, claiming in 1837, for example, that ice sheets had once extended over Europe as far south as the Mediterranean. But there were no signs of glacial erratics or 'drift', the deposits left by retreating glaciers, and he was obviously wrong. Despite widespread interest in his glacial theory, Agassiz's wilder claims encountered almost universal scepticism. Twenty years passed before the notion of an ice age became mainstream science, largely as a result of numerous discoveries of relatively recent glaciation, many of them by Scandinavian, Scottish and Swiss geologists, who lived in recently or still glaciated areas. Agassiz's Ice Age theory also attracted widespread attention on the other side of the Atlantic. As early as 1839, palaeontologist Timothy Conrad reported signs of glacial activity in western New York State, the first of many such observations. In 1852, a scientific expedition showed that a huge ice sheet covered Greenland. Late 19th-century explorers established the extent of the Antarctic ice sheet.

Gradually, enough evidence of glaciation and its related phenomena accumulated to allow the first comparisons between modern ice sheets and Ice Age conditions. By the 1860s, the glacial theory was well established on both sides of the Atlantic, except for scattered and persistent opposition, much of it from religious bigots.

Agassiz visited the United States in 1846 at the urging of Charles Lyell, to see evidence of Ice Age glaciation in North America for himself. He said in a later lecture: '[around Boston] I was met by the familiar signs, the polished surfaces, the furrows and scratches, the line engravings of the glacier ... Here also this great agent had been at work.' He travelled widely after accepting a professorship at Harvard

University in 1847 and became a powerful advocate for all kinds of natural science. He did little more Ice Age research, but in 1865 he found traces of glaciation in the Andes, and declared that North American ice sheets had expanded far into the south, even through the torrid Amazon Basin. 'Agassiz has gone wild about glaciers', complained Charles Lyell, and with good reason. By the time of his death in 1873, Louis Agassiz had spent decades in the public eye as an advocate for science, even if his ideas were often outdated. He was, for example, opposed to the theory of evolution and believed that the Divine Hand had crafted the Ice Age world. For all his wild ideas, however, Agassiz was one of the key founders of Ice Age science.

Humans and ice sheets

Like all major scientific breakthroughs, the credit for the discovery of the Ice Age should not go to Agassiz alone. He developed his theory of a great Ice Age during a period of profound change in many scientific disciplines. Stratigraphic geology – the study of layers of rock – made Agassiz's observations, and those of his contemporaries, possible. Charles Darwin developed his theory of evolution and natural selection during the years of ferment over the Ice Age, publishing *On the Origin of Species* in 1859. This period also saw the final and conclusive proof that humans had lived at the same time as long-extinct animals, during a geological era that extended back far longer than the mere 6,000 years of biblical chronology. A primitive-looking Neanderthal skull had come to light in a German cave in 1856, seven years before the biologist Thomas Henry Huxley established the close

This well-preserved Neanderthal skull was found in Germany's Neander Valley in 1856 and, along with the publication of Darwin's seminal work *On the Origin of Species* three years later, helped to revolutionize ideas about the age and origin of humanity. Neanderthal fossils are often associated with cold-adapted species such as reindeer and woolly mammoth, demonstrating that these hominins lived in Europe during glacial periods.

An Ice Age colossus. This mammoth (*Mammuthus primigenius*), originally found preserved in the Siberian permafrost, was excavated at Yakutia in 1903 and exhibited (stuffed) at the St Petersburg Museum of Zoology. During the 19th and early 20th centuries mammoths were a subject of great wonder, as well as a major source of ivory in Russia.

anatomical relationships between humans and their closest living relatives, the chimpanzees. In 1862, labourers digging the foundations for a railway station at Les Eyzies in southwestern France unearthed burials of fully modern humans in the same layers as cold-loving animals like reindeer. Here was definitive proof that humans had lived in Europe during a period of extreme cold – Louis Agassiz's Ice Age.

A cartoon of a crowded 19th-century lecture on prehistory, parodying the relationship between humans and their fossil ancestors. The idea that humans had evolved from apes was met with a huge amount of opposition in the 19th century, in particular from some members of the clergy.

The discovery of the Ice Age was one of the great achievements of 19th-century science, resulting not only from the inspired researches of a single geologist, but also from sustained observations all over the world. The Discourse of Neuchâtel coincided with an explosion of geological research triggered by the Industrial Revolution and its insatiable demand for minerals, as well as by a growing scientific curiosity about the earth and its history. The great scientific expeditions that explored and mapped the American West during the 1870s were part of the new geology. So were numerous expeditions that took fieldworkers onto glaciers and high on mountain slopes, where they studied the workings of glaciers – the ways in which they expanded and shrank, the formation of their ice from layers of tightly packed snow, and the distinctive 'tills' of rock and other debris deposited by advancing and retreating ice sheets. By studying the extent of such geological layers, Victorian geologists were able to map the extent of Ice Age glaciation. In Europe, ice sheets had mantled Scandinavia and the Alps, while in North America vast ice sheets had once extended from New York State across to the Seattle region of Washington State. By 1875 a worldwide map showed that Ice Age glaciers and ice sheets had covered 44 million sq. km (17 million sq. miles), well over three times today's coverage. Most Ice Age glaciation, some 26 million sq. km (10 million sq. miles) of it, lay in the Northern Hemisphere. For tens of thousands of years, the world had been a very different place, much of it covered by vast ice sheets and suffering under bitter cold.

The new map raised numerous questions. When had the Ice Age begun and how long had it lasted? When had it ended? Had there been more than one cold period during the millennia – times when ice sheets expanded, before contracting again? And why had the world suddenly entered an ice age after millions of years of tropical climate? These questions have preoccupied scientists ever since, and continue to do so. Louis Agassiz's *Eiszeit,* the geologist's Pleistocene, has acquired a formidable complexity.

right This map by the influential American geologist Thomas C. Chamberlin (1843–1928) shows the maximum extent of the Laurentide and Cordilleran ice sheets over North America during the Ice Age. Chamberlin's surveys of glacial deposits in the northern United States led him to propose a theory of multiple glaciations that is still relevant today. As well as contributing hugely to Ice Age science, Chamberlin was one of the first to emphasize the role of carbon dioxide as a major regulator of the earth's temperature.

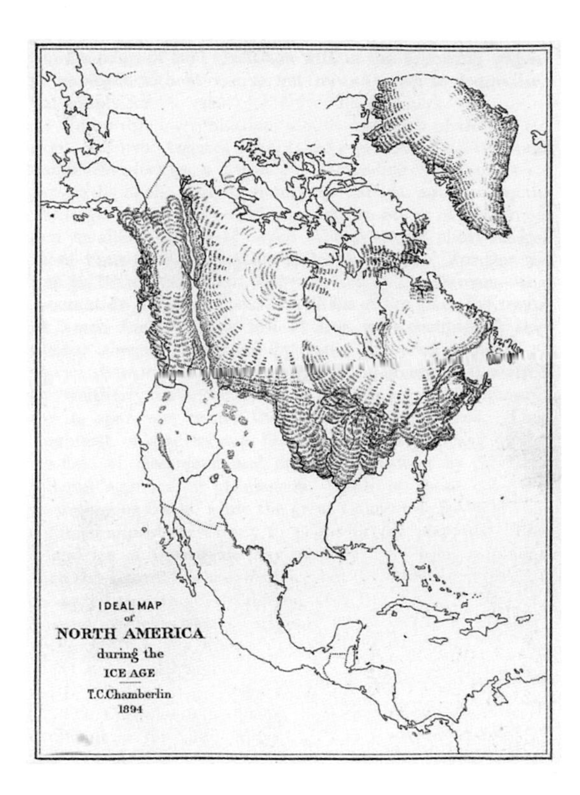

below Although born, as he said, 'on a moraine' in southeastern Illinois, Chamberlin moved to southern Wisconsin as a child, and in 1873 began work on a comprehensive geological survey of the state. This survey and others resulted in detailed maps of the glacial deposits of North America such as the one shown below, and hugely improved knowledge of the North American Pleistocene.

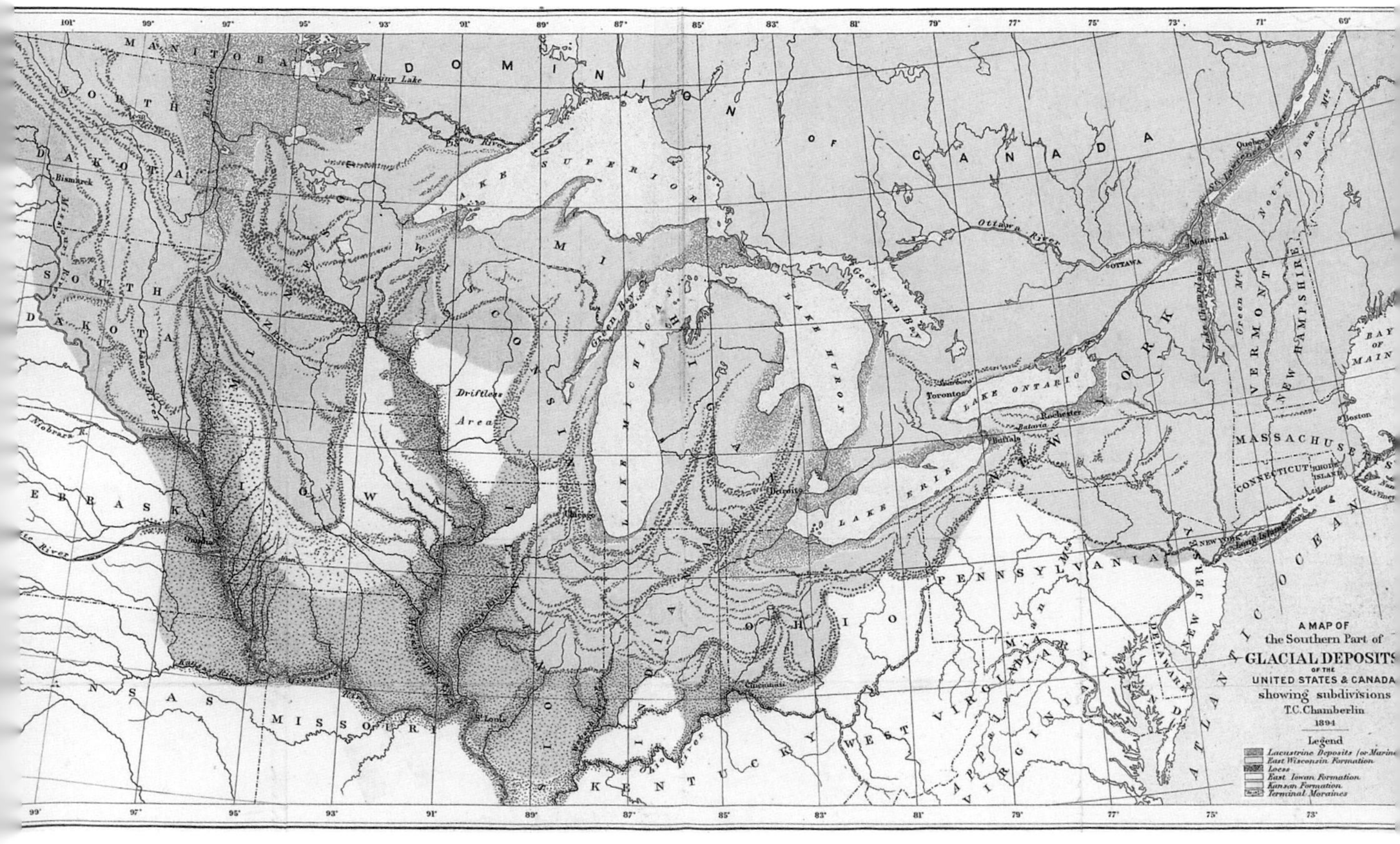

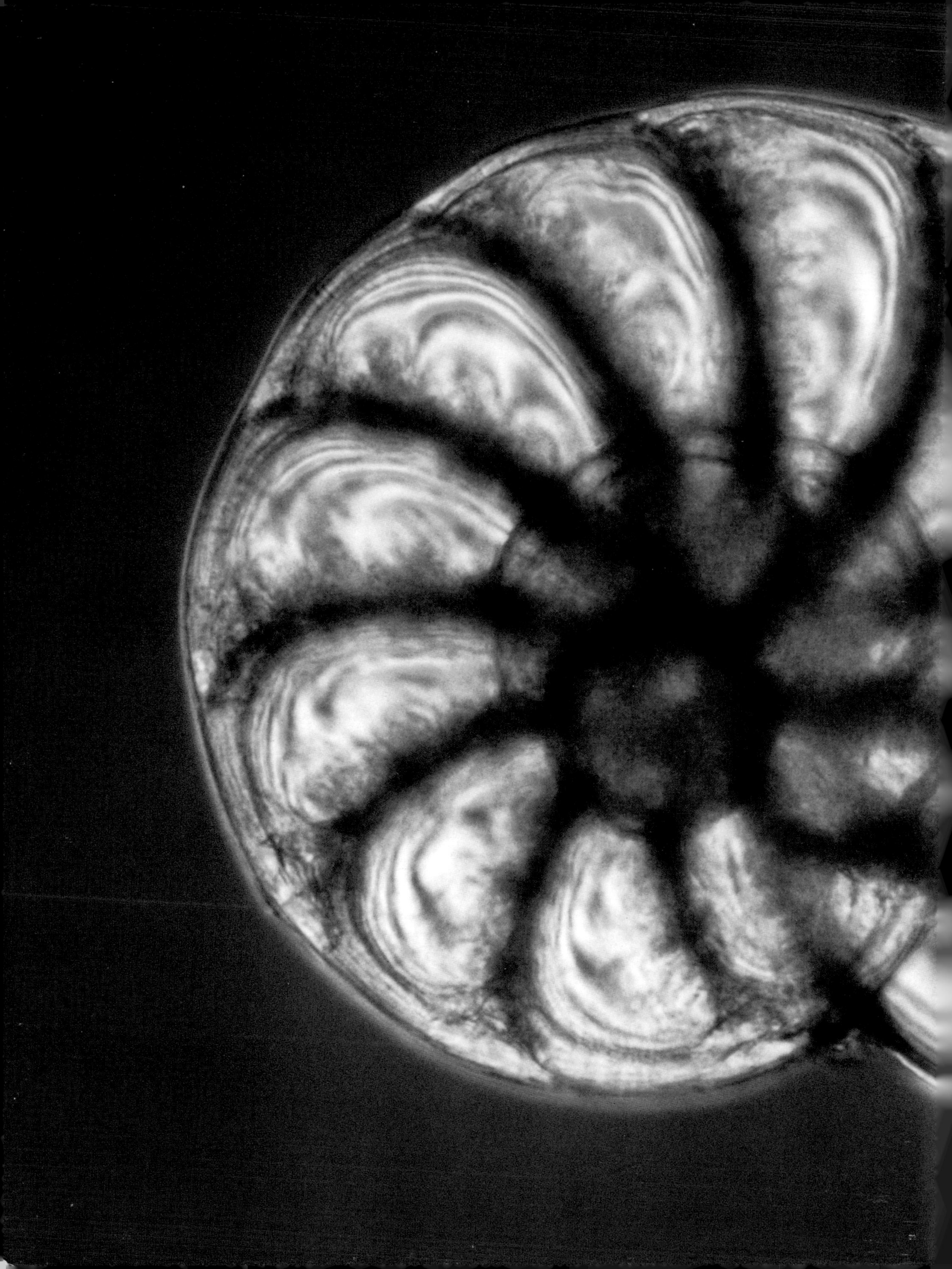

2
Searching for Clues

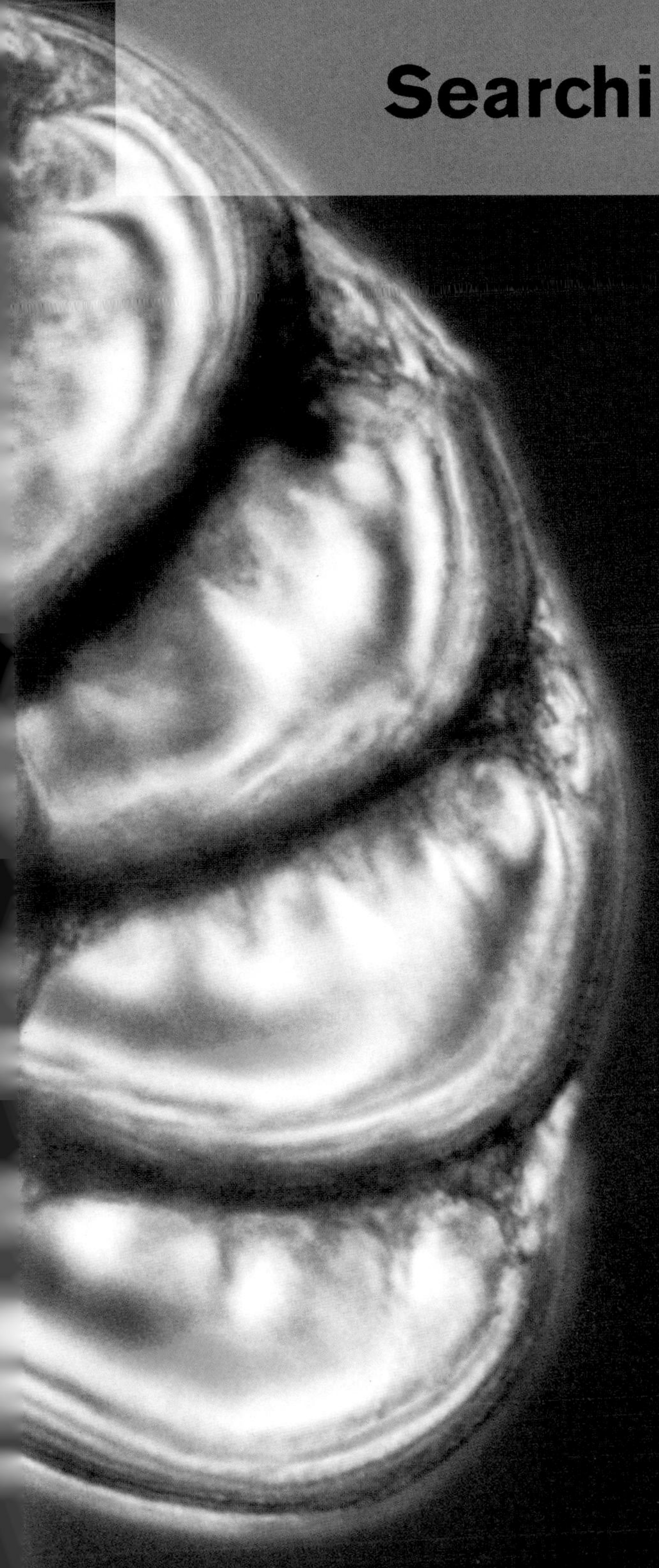

IN THE 19TH CENTURY, Louis Agassiz had developed a relatively simple vision of an Ice Age world covered in massive ice sheets. He wrote in *Geological Sketches* (1866) 'The long summer was over ... For ages a tropical climate had prevailed ... and ... giant quadrupeds ... had possessed the earth. But their reign was over. A sudden intense winter, which was to last for ages, fell upon our globe.' His image, however, did not survive long.

Agassiz's successors quickly learned a great deal about the workings of glaciers and how they flowed. Their industrious mappings soon yielded some surprises that contradicted Agassiz's original vision of a huge ice sheet that covered most of the Northern Hemisphere from a centre at the North Pole. The massive Laurentide ice sheet, which covered most of Canada from a centre on Hudson Bay, had both a southern and northern boundary. In the Southern Hemisphere, the 13 million-sq. km (5 million-sq. mile) Antarctic ice sheet expanded only slightly during the Ice Age. All other southern ice sheets were in mountainous areas such as the Andes. But where had all the water tied up in the great glaciers and ice sheets come from?

previous pages A light micrograph of a shelled protozoan, *Nonionina depressula*, belonging to the order Foraminifera. These creatures, also known as forams, are mostly marine; some are sea-bottom dwellers while others form part of the surface plankton. Although tiny, they are in fact one of our greatest sources of information about the Ice Age.

Sea levels and loess

As early as 1841, in his article *The Glacial Theory of Prof. Agassiz*, Scottish geologist Charles Maclaren estimated that 'the abstraction of the water necessary to form the said coat of ice would depress the ocean about 800 feet [240 m]'. Few people took him seriously. By 1868, the long-vanished ice sheets were better known, so geologist Charles Whittesey from Cleveland, Ohio, used calculations of continental ice

Tall prairie grass grows on the loess hills of western Iowa. Formed during dry glacial periods out of millions of tons of wind-blown silt, weathered loess soils provide fertile soils for modern-day farmers.

thickness to estimate that the oceans had been 107–122 m (350–400 ft) below modern levels, enough for large continental shelves off major land masses to have been exposed. Alaska and Siberia would have been one, and England part of the Continent.

Ice sheets and lower sea levels: the Ice Age world was very different from today. But what about the effects of intense cold and widespread glaciation on areas adjacent to the ice sheets? While searching for glacial moraines, geologists discovered that 2.6 million sq. km (over 1 million sq. miles) of Europe, Asia and North America were mantled in a layer of fine, uniform silt up to 3 m (10 ft) thick. Borrowing a term from Swiss-German farmers, they named it loess (*lösch*, 'loose'). The minute silt grains were uniform but angular, often stratified in layers and sometimes found in isolated patches. In 1870, German geologist Ferdinand von Richthofen identified loess as windblown dust. 'There is but one great class of agencies which can be called on for explaining the covering of hundreds of square miles ... with a perfectly homogeneous soil', he wrote in *Geological Magazine*. 'Whenever dust is carried away by *wind* from a dry place, and deposited on a spot which is covered by vegetation, it finds a resting place.' He theorized that great clouds of fine silt from glacial streams were blown away by high winds blowing off the glaciers. It was no coincidence that Europe's first farmers of 8,000 years ago, who had no ploughs, favoured loess, also a vital ingredient in the soils of the American farm belt.

The legacy of the Ice Age can be seen throughout the world. On the banks of the Huang He River Valley in northern China terraces have been cut into the vast loess deposits in order to grow wheat.

As Victorian geologists looked more closely at glacial deposits and shorelines in both Europe and North America, they found themselves mapping not one Ice Age, but a series of glacial events, separated by periods of much warmer climate. They realized that the Ice Age was a long epoch of constantly fluctuating glacials and warmer interglacials. Just how many remained a mystery, and even today many details of these cycles are still unknown.

Why an ice age?

The early geologists also puzzled over another major question: why did ice sheets advance to cover almost a third of the world's continents, then retreat again before growing once more? Theories abounded, revolving around such factors as changes in solar energy and sunspot activity, shifting distributions of dust particles in space, and varying concentrations of carbon dioxide in the atmosphere. One dramatic hypothesis, proposed by New Zealander Alex T. Wilson as recently as 1964, suggested that a collapse of large segments of the Antarctic ice sheet into the ocean, as a result of the increasing weight of the ice caused by snow accumulation, would lead to an ice age. Surrounding water would be covered with reflective layers of floating ice that would send back more and more of the sun's radiation into space, thereby causing an ice age. However, such glacial surges, well documented in some mountain glaciers, would have sea level rises occur *before* an ice age, not during it, as actually happened.

Could dust from periods of intense volcanic activity have triggered an ice age? Intense eruptions could have released vast quantities of fine dust into the atmosphere, masking the earth from the sun and causing global temperatures to drop. We know that volcanic eruptions can cause cold temperatures. After Mount Tambora in Southeast Asia exploded in 1815, Europe experienced the 'year without a summer' in 1816. Conditions in Switzerland were so cold that crops failed and people starved. Mary Shelley, on holiday there, stayed indoors and invented the story of Frankenstein. In 1883, Krakatoa Island blew into space with a bang that was heard 4,828 km (3,000 miles)

The Krakatoa eruption of 1883 caused widespread tsunamis and huge casualties. Dense ash clouds rose high into the atmosphere and created spectacular sunsets over a wide area of the world. It has been proposed that the Ice Age could have been triggered by dust from a series of great eruptions blocking out the sun's rays, but no evidence for such volcanic activity has ever been found.

away. Red sunsets were seen around the world for two years and global temperatures dropped significantly until the dust particles drifted back to earth and the climate returned to normal. These were isolated eruptions. What would have happened if there had been sustained volcanic activity with correspondingly prolonged temperature drops? But no one has yet found geological evidence for such periods at the onset of the Ice Age.

So far, despite nearly two centuries of diligent field research and theorizing, no one has come up with a convincing geological theory for ice ages. The most promising explanations come from a body of theory that emerged only five years after Louis Agassiz gave his discourse on the Ice Age at Neuchâtel.

In 1842, a French mathematician, Joseph Alphonse Adhémar, published *Revolutions of the Sea*, a book in which he argued that ice ages resulted from variations in the ways the earth moved around the sun. He theorized that glacial climates occurred as a result of shifts in the equinoxes along the earth's orbit, a cycle completed every 22,000 years. Adhémar's contemporaries rejected his ideas as mere fantasy, especially his scenario of a melting Antarctic ice sheet causing ice-laden tidal waves that flooded land in the Northern Hemisphere previously drained by the ice sheet's formation. However, the astronomy was another matter – the first astronomical theory for the Ice Age.

James Croll (1821–1891) was a self-taught Scottish scientist, with remarkable intellectual powers. His theory that variations in the earth's orbit had a profound effect on global climate predicted that there were multiple Ice Age glaciations.

Adhémar's mantle passed to a self-taught Scot, James Croll, who dropped out of school at the age of 13, but continued his own education. He became a millwright in the hope that the occupation would be congenial for someone of his theoretical bent. He then became a carpenter before working in a tea room, running a store, a hotel, and selling insurance, but soon found that his abstract mind was useless in the practical world. In 1859 he became a janitor in the Andersonian Museum in Glasgow, which boasted of a fine scientific library. Croll could now indulge what he called an 'almost irresistible propensity towards study'. He began with physics, then turned to geology and the Ice Age in 1864.

By this time, Croll was aware of Adhémar's book, and also of the research of the French astronomer Urbain Le Verrier, who had shown that the elongation of the earth's orbit was changing slowly, but continually. This, Croll believed, was the real cause of the Ice Age. He published a widely praised paper on the subject in the *Philosophical Magazine* for August 1864, then mastered the complex mathematics behind Le Verrier's ten years of orbital calculations. With single-minded intensity,

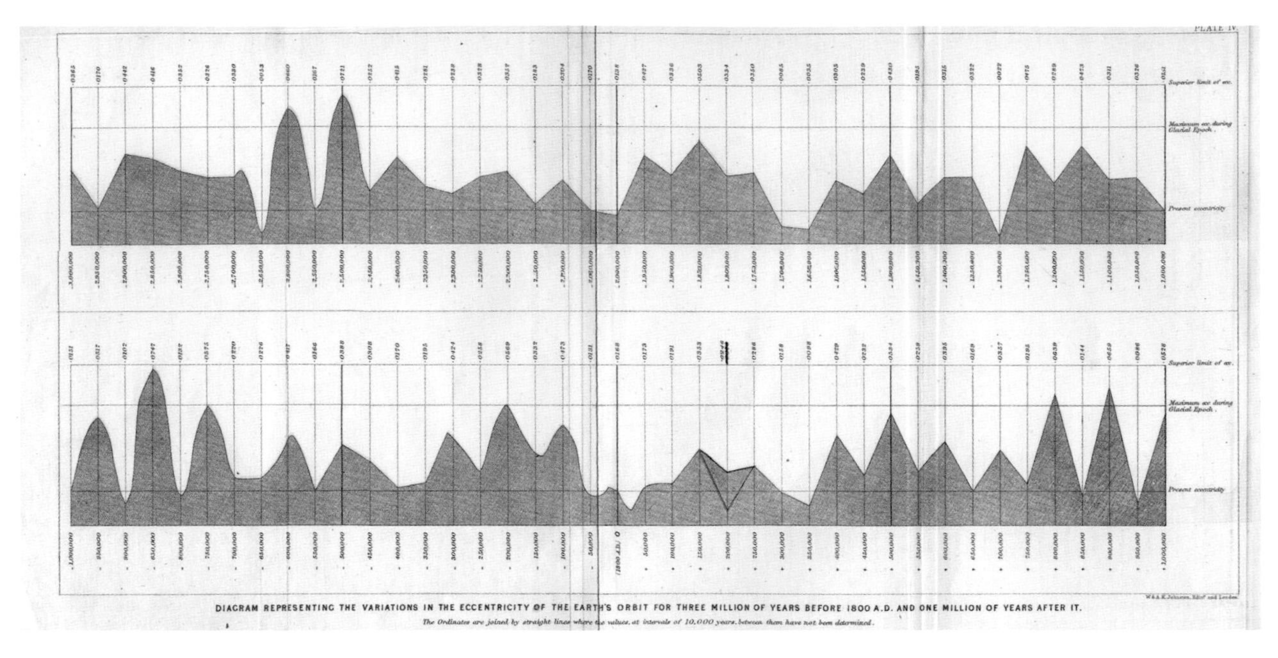

Croll plotted the changes in orbital eccentricity over the past 3 million years and discovered that it changed cyclically, alternating between long periods of low and high eccentricity. He believed that ice ages were caused by changes in the distance between the earth and the sun, as measured on 21 December. When this distance exceeded a critical value, winters in the Northern Hemisphere were cold enough to trigger an ice age in that part of the globe. In 1875, Croll published *Climate and Time*, which summarized his theory. He was elected a Fellow of the Royal Society for his research.

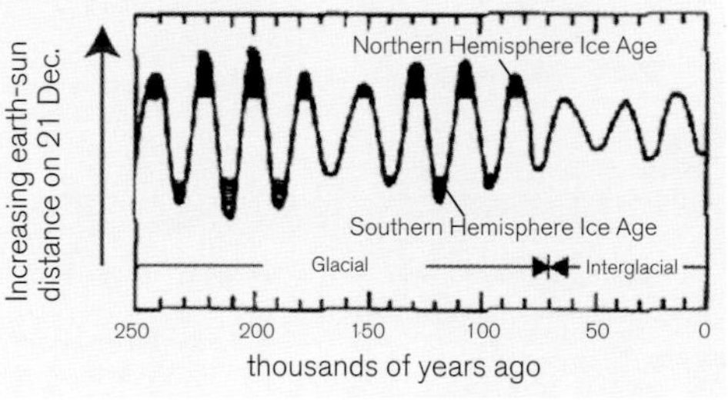

above and top Croll's diagrams demonstrating his theory of Ice Age glaciation. This was based on an assumption that ice ages were caused by changes in the distance between the earth and the sun as measured at the winter solstice. He believed that when this distance reached a critical point, Northern Hemisphere winters were cold enough to cause an ice age.

opposite The earth's orbit around the sun oscillates between more and less elliptical states over a 96,000-year cycle (top); the tilt of the earth's axis fluctuates between 21.5 and 24.5 degrees every 41,000 years (lower left); and the earth 'wobbles' as it spins, like a revolving gyroscope. Each 'wobble' lasts 21,000 years (lower right).

The Croll theory of the Ice Age was enormously influential, especially after the Scottish geologist James Geikie championed Croll's ideas in his book, *The Great Ice Age* in 1874, the first full-length treatment of the subject since Agassiz's 1840 *Studies on Glaciers*. There were, however, serious weaknesses in Croll's theory. He believed the last Ice Age ended with a change in eccentricity about 80,000 years ago. At the time, no one could date geological strata, so there was no means of validating his chronology. Eventually, American geologists used data from Niagara Falls and the Falls of St Anthony on the Mississippi River near Minneapolis to argue for an end date as recent as 15,000 to 10,000 years ago. At the same time, meteorologists raised serious objections, arguing that Croll's variations in solar heating were too small to have any effect on global climate.

By 1894, Croll's theory was little more than a historical curiosity, regarded as invalid and almost forgotten, until a Serbian engineer became 'attracted to infinity' and said: 'I want to grasp the entire universe and spread light into its farthest corners.'

The Milankovitch curve

The engineer was Milutin Milankovitch (1879–1958), who later became Professor of Applied Mathematics at the University of Belgrade. Still under the spell of infinity, he immersed himself in the challenge of creating a mathematical theory that would explain ancient and modern climates, not only on earth but on Mars and Venus as well. Unlike Adhémar and Croll, Milankovitch had the mathematical training to calculate the magnitude of orbital variations with considerable accuracy.

For 30 years, Milankovitch laboured on his theory, even when travelling or on holiday, insisting that a desk be provided in his hotel rooms. After prolonged labours – and this was, of course, in the days before computers – he found that a decrease in the earth's axial tilt caused a fall in summer radiation. However, a lessening of the distance between the earth and sun at any season increased radiation. The strength of these effects varied with latitude: the radiation curves for high latitudes were dominated by the 41,000-year oscillation of the earth–sun distance tilt cycle, the cycle at the Equator being 22,000 years. Using data on mountain snow lines, he was able to determine how much increase in snow cover would result from a given change in summer radiation (see Chapter 4).

After writing a book on his theory in 1920, Milankovitch began collaborating with the well-known German climatologist Wladimir Köppen and his son-in-law, the geologist Alfred Wegener. With their encouragement, he calculated the summer

Milutin Milankovitch (1879–1958) in a 1943 portrait by the Serbian painter Paja Jovanovic (1859–1957).

below right Deposits on ocean floors can be used to reconstruct past climatic change through the study of the chemical signatures preserved in the shells of fossilized micro-organisms. This deep-sea core record shows the major temperature fluctuations of the past 6 million years.

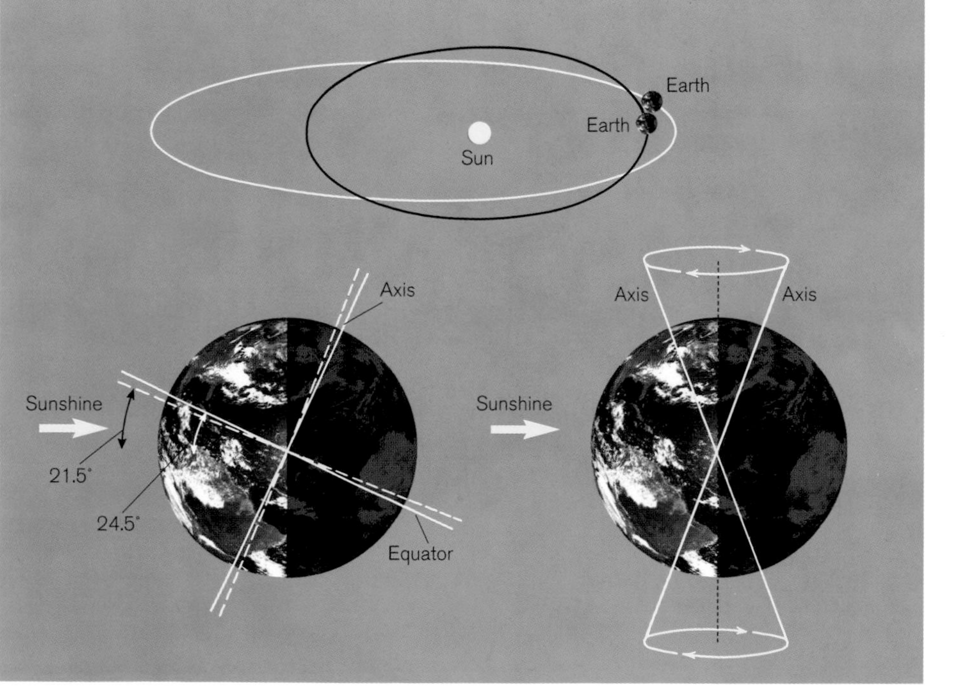

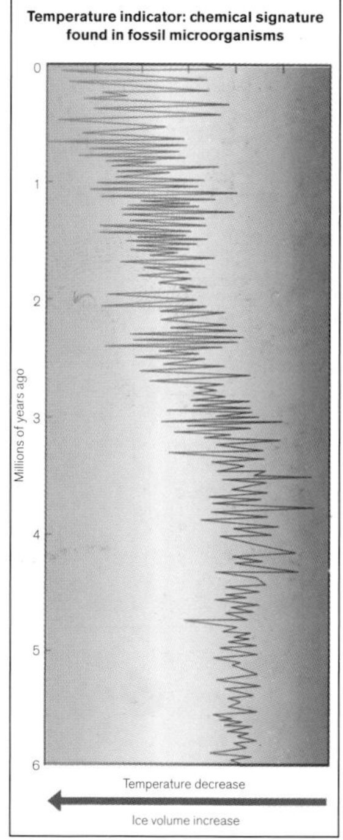

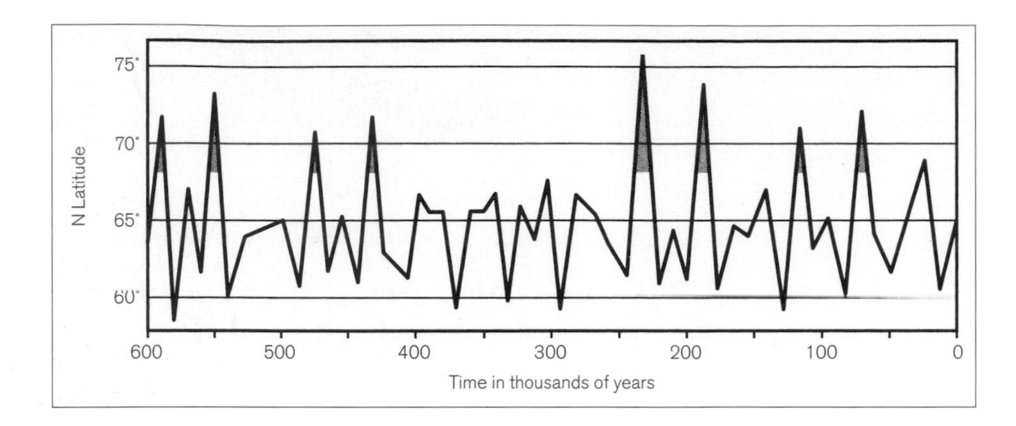

Köppen and Wegener used this graph – devised by Milutin Milankovitch – in their widely read book *Climates of the Geological Past*. It shows the 'equivalent latitude' for 65°N in summer over the past 600,000 years, based on the amount of solar energy received (insolation). Therefore a peak on the graph of 75°N indicates that during the summer of that year the earth only received as much insolation at 65°N as it does today at 75°N. The shaded peaks on the graph indicate periods when the equivalent latitude was higher than today – a factor Milankovitch believed would lead to glacial periods.

variations in solar radiation at three northern latitudes over the past 650,000 years. The calculations took him 'a full hundred days from morning until night' and resulted in the famous Milankovitch curves; the curve revealed low points that appeared to coincide with four Ice-Age glaciations recently identified in the Alps. In all, there were nine radiation dips indicating glaciations on the Milankovitch curve, forming a distinctive, irregular pattern. The radiation curve appeared in Köppen and Wegener's widely read *Climates of the Geological Past* in 1924.

There was by now glaciological data to compare to the curve. In 1909, Austrian geologists Albrecht Penck and Eduard Brückner had published a classic monograph on Ice Age glaciation in the Alps. *Die Alpen im Eiszeitalter* described four glacial periods – Günz, Mindel, Riss and Würm – when the Alpine snow level was an estimated 1,000 m (3,280 ft) lower than today. Penck and Brückner's four glaciations appeared to coincide well with Milankovitch's graph. Their somewhat crude observations soon gave way to more refined schemes, notably by German geologists Barthel Eberl and Wolfgang Soergel, which appeared to match the curve even more closely.

As soon as the Milankovitch curves appeared, geologists began matching them to glacial deposits in the field. American geologists Thomas C. Chamberlin and Frank Leverett identified four ice sheets that had once mantled much of North America, the timing of which might match those in the Alps, but their data were criticized as being too general for any accurate correlations.

During the 1930s and 1940s, in the days before the radiocarbon revolution, the Milankovitch calendar became the yardstick for Ice Age chronology in Europe, although American scholars, far from the Alps, were more sceptical. There were other critics, too, among them the German geologist Ingo Schaefer, who unearthed warm-loving molluscs in Alpine glacial deposits. The major objections came from meteorologists, who pointed out that Milankovitch had ignored the role played by the atmosphere and the ocean in transporting heat. Milankovitch swept these criticisms aside. 'I do not consider it my duty to give an elementary education to the ignorant,' he remarked. So strong was the support for his ideas that the Milankovitch time scale became the only viable chronology for the Ice Age during the 1930s and 1940s. Then, suddenly, support for the curve collapsed like a house of cards, when the development of the radiocarbon and potassium-argon dating methods completely transformed Ice Age chronology.

Willard Libby (1908–1980), Nobel laureate and father of radiocarbon dating.

The dating revolution

Willard Libby, a University of Chicago chemist, had cut his scientific teeth working on the Manhattan Project during World War II. During three years of research, starting in 1946, he discovered that a radioactive form of carbon (carbon-14 or radiocarbon) is produced in small quantities in the atmosphere. Eventually the radioactive atoms in the atmosphere are absorbed into the bodies of living animals and plants. Once they die, the atoms in their organic tissues begin to decay and slowly turn into inert nitrogen at a rate that can be measured with a geiger counter. Convinced that he had found a way of dating the past, Libby began measuring the proportion of carbon atoms in ancient organic materials like wood, hearth charcoal and even bone. He tested the method extensively, found that it worked, but also discovered it could only

A scientist takes a sample of shavings from a reindeer bone for radiocarbon dating. The procedure measures the ratio between the radioactive isotope carbon-14 (14C) and the stable carbon-12 (12C) isotope in a sample. The ratio of 14C to 12C in the sample may be related to the time since the death of the animal or plant being investigated. Radiocarbon dates are calibrated into calendar dates using tree rings, ice cores and other sources.

be used to date fossils less than 40,000 years old. Nevertheless, the new radiocarbon dating method could be used to date the closing millennia of the Ice Age.

One early radiocarbon user was American geologist Richard Foster Flint, who collected large numbers of samples from late Ice Age glacial deposits in the eastern and central United States. He soon found that he was dealing not with one glaciation, but with two, the later of which culminated about 18,000 years before the present, and was in full retreat before 10,000 years ago. It soon became clear that Milankovitch's astronomical chronology for the late Ice Age was hopelessly inaccurate. Radiocarbon dates were so accurate that a geologist could collect samples from deposits left by retreating ice sheets and date the various stages of their retreats. Thus began a revolution in late Ice Age chronology, which continues to this day, as geologists and climatologists puzzle out minor details of climate change over 40 millennia.

Forty thousand years is but a metaphorical blink of an eye when stacked against the enormous span of geological time. How, then, could one date earlier Ice Age glaciations and interglacials more accurately than Milankovitch? Once again, science provided the answer in the late 1950s – potassium-argon dating.

In 1959, Mary Leakey unearthed *Zinjanthropus*, a robustly built hominin, in an ancient lake bed at Olduvai Gorge in Tanzania. She called her discovery 'Dear Boy', and it was to be a find that helped revolutionize the understanding of human

Louis Leakey (1903–1972) and Mary Leakey (1913–1996) displaying the skull of *Zinjanthropus*, a robust Australopithecus unearthed by Mary at Olduvai Gorge, Tanzania, in 1959.

Dating the remote past using an electronic furnace in a potassium-argon laboratory at the University of California, Berkeley. This dating method revolutionized the chronology of the Ice Age and early human evolution.

evolution. At the time, except for intelligent guesswork, no one had any idea how old humanity was, but it was often assumed to be in the region of 250,000 years. A year later, Louis and Mary Leakey found a more gracile hominin, *Homo habilis* ('Handy Person'), also in a stratified lake bed associated with volcanic deposits, but underlying the *Zinjanthropus* site, and thus somewhat earlier. The dating of these remarkable finds became an urgent priority.

Fortunately, the Olduvai discoveries coincided with research by a group of scientists at the University of California, Berkeley, who developed a geochronological method known as potassium-argon dating. They based their new dating technique on the radioactive decay of potassium, a common element found in volcanic rocks. The 40 Ar argon isotope has a much slower decay rate than radiocarbon and can be used to date rocks billions of years old. Almost immediately, the potassium-argon researchers turned their attention to Olduvai Gorge, which lies in an area of ancient volcanic activity. They soon dated *Zinjanthropus* and *Homo habilis* to *c.* 1.75 million and 2 million years ago respectively, a new chronology for human evolution that caused a sensation.

Since then, potassium-argon dating has carried the frontiers of human evolution back even further to at least 4.5 million years ago. And potassium-argon dates for the Ice Age itself coincided with a period of gradual cooling about 2.5 million years ago. So early human evolution extends beyond the earliest millennia of the Ice Age.

Coring the oceans

The Ice Age may have been about 1.5 million years long, but what about the pattern of glacials and interglacials over this long time frame? Glacial deposits on land are not necessarily the most accurate barometers of Ice Age climate change. So climatologists and oceanographers turned to the sea bed instead.

The idea of looking for Ice Age climate change in the ocean is nothing new. 'In the deep recesses of the sea, buried under hundreds of feet of sand, mud, and gravel, lie multitudes of the plants and animals, which ... were carried down by rivers into the sea,' wrote James Croll in a moment of inspired and prophetic speculation in his book *Climate and Time*. In Croll's day, the oceans were a mystery, until the researches of scientists aboard HMS *Challenger* investigated ocean floor sediments in 1872–75. Away from coastlines and continental shelves, the *Challenger* researchers collected samples of fine-textured oozes from relatively shallow tropical and temperate waters that contained the mineralized remains of Foraminifera ('forams' for short), planktonic animals. The *Challenger* researchers discovered that some forams and other planktonic organisms lived only in cold water, while others dwelt in warmer waters. Here, at least theoretically, was a way of reconstructing the changing temperatures of the Ice Age – by recovering stratified columns of deep-water sediments and analysing the forams in them.

Unfortunately, no one had devised a way of recovering deep-sea cores without distorting the fine deposits, so research languished. Then, in 1947, a Swedish oceanographer, Björe Kullenberg, developed a piston borer that sucked sediments into a tube and regularly produced cores up to 15 m (50 ft) long. A series of cores from the Swedish Deep Sea Expedition of 1947–48 produced samples from both the Atlantic and the Pacific oceans. Soon a classic monograph on 39 Atlantic cores showed that they contained evidence for at least nine glacial periods during the Ice Age.

While this research was under way, another oceanographer, Cesare Emiliani of the University of Chicago, was examining the isotopic composition of oxygen atoms in fossil forams. His results compared well with those for recent millennia obtained from the same cores at Lamont, but were very different for earlier times. Meanwhile, David Ericson of Lamont had used more than 3,000 cores to document a sequence of at least nine glacial periods. Eight cores from early in the Ice Age showed a

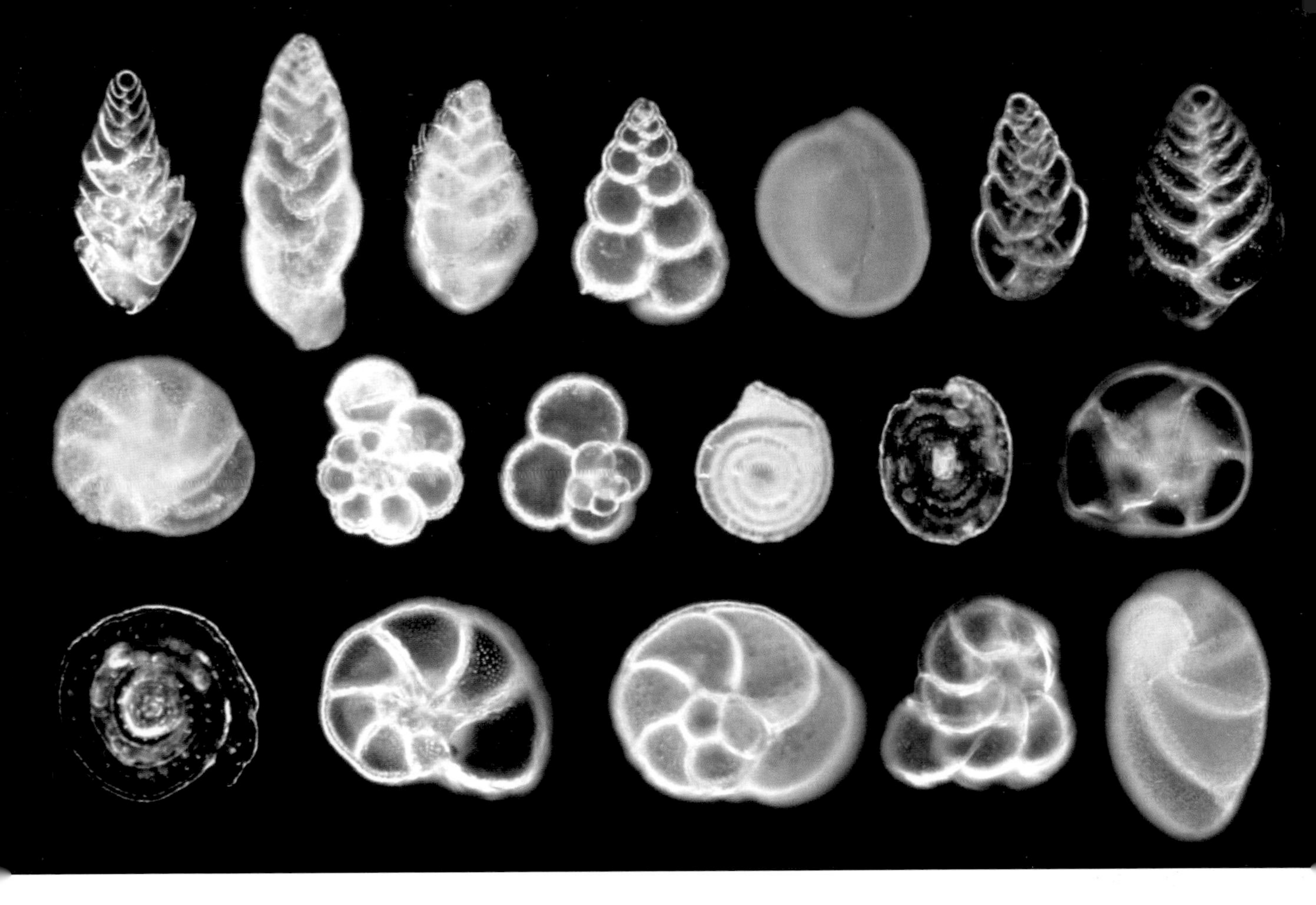

A light micrograph of assorted species of shells belonging to the order Foraminifera. Foraminiferan shells form an important component of chalk and of many deep-sea oozes. In previous geological ages Foraminifera occurred in such enormous numbers that their shells, largely of calcium carbonate, have formed immense fossil deposits seen today as limestone.

boundary where star-shaped fossil plants called 'discoasters' became extinct. This event occurred an estimated 1.5 million years ago.

The Ericson chronology was controversial, largely because no one had developed an accurate way of dating the molluscs in the cores. Simultaneously, Emiliani worked with teams of researchers at Chicago, who were using oxygen atoms to study ancient sea temperatures. Seawater contains both oxygen-18 and oxygen-16 isotopes, the former being heavier than the latter. Both are present in the calcium carbonate skeletons of marine organisms. The amount of the heavier isotope that an animal extracts from the surrounding water depends on the water temperature, so the skeletons of organisms that thrived in cold water would have heavier concentrations of oxygen-18. The American geochemist Harold Urey had argued that it would be possible to calculate water temperatures using the ratios between the two isotopes in fossil skeletons. Emiliani applied what was now called the 'isotopic thermometer' to marine Foraminifera from the Ice Age. By 1955, he had analysed eight deep-sea cores and worked out a temperature curve for the Atlantic and Caribbean that documented seven glacial periods over the last 300,000 years, and a temperature decline of about 6°C (11°F) during a glacial period in the Caribbean. He also pointed out that his chronology corresponded fairly well with the Milankovitch curve.

Eventually, statistical analyses of the foram assemblages showed that the variations in isotopic readings resulted from changes in the volume of Ice Age ice sheets, not from temperature changes. This approach allowed researchers to use the isotope techniques to measure the volume of ice and statistical methods to record ocean temperature changes.

Magnetic reversals and the Ice Age boundary

When had the Ice Age begun? Deep-sea cores estimated about 1.5 million years ago, but an entirely different approach, palaeomagnetism, produced a different estimate.

In 1906, a French geophysicist, Bernard Brunhes, who was investigating the earth's magnetic field, discovered that as a newly baked brick cools, it becomes slightly magnetized. He also discovered that cooling volcanic lava behaves like bricks. He soon found that the direction of magnetism in some lava flows was the direct opposite of that of the earth's present magnetic field. Over 20 years later, a Japanese geophysicist, Motonori Matuyama, became convinced that the earth's magnetic field had reversed itself many times during geological time and at least once during the Pleistocene. Brunhes' and Matuyama's findings were not confirmed

left Aurora borealis, a natural light display in the ionosphere, usually visible in northern latitudes. Aurorae are the visible manifestation of the earth's magnetic field. They are caused by collisions of charged particles from the earth's magnetosphere which connect with the earth's magnetic field and excite atoms and molecules in the upper atmosphere.

opposite A computer depiction of the earth's magnetic field. The blue lines represent the magnetic field lines, which extend from the magnetic poles. These are near, but not at, the North and South Poles. The magnetic field is thought to be generated by the movement of molten iron in the earth's core. It protects the earth from high-energy radiation from the sun. The magnetic field is not stable and has reversed abruptly and irregularly during the earth's history, most recently about 780,000 years ago, at what is known as the Brunhes–Matuyama Event. These reversals can be seen in deep-sea cores.

until 1963, when a team of American scientists proved that the magnetic field reversals were global events. There were two during the Ice Age – the Brunhes–Matuyama Event of 780,000 years ago, and the Olduvai Event, dating to about 1.8 million years ago. Soon the same reversals were identified in Antarctic deep-sea cores.

Palaeomagnetism helped provide a more refined date for the beginning of the Ice Age. Scientists at the Woods Hole Oceanographic Institution, USA, showed that the first appearance of cold-loving species was during the Olduvai Event, some 1.8 million years ago.

Climatic cycles

Meanwhile, intensive research on both land and the sea bed was changing perceptions of Ice Age climate. In the late 1960s, Wallace Broecker and Jan van Donk at the Lamont-Doherty Earth Observatory used isotopic measurements of forams from Caribbean core V12-122 to conclude that the major cycle in the isotopic record recorded by Ericson and Emiliani was 100,000 years. This climatic cycle was notably asymmetrical, with a slow cooling and a warming so rapid that they named it a 'termination'. So the major Ice Age glaciations were spaced about 100,000 years apart, a cycle confirmed not only by core research but also from land-based loess deposits in Czechoslovakian brickyards. Almost simultaneously, William Ruddiman and Andrew Macintyre, also at Lamont, used Atlantic sea cores on a north–south line to show that the Gulf Stream had moved south during glacial periods to flow eastwards towards Spain during eight climatic cycles since the Brunhes–Matuyama Event. So the Gulf Stream marched to the same climatic drum beat.

By 1971, there was enough research activity for a truly international effort. This took the form of the CLIMAP project, designed to map the surface of the earth during the Ice Age and to study Ice Age climate change. In 1973, the scope of the project widened to include studies of the cycles of Ice Age climate, which culminated in the publication of a global map of ocean temperatures and glacier distribution 18,000 years ago, at the height of the late Ice Age. However, the project lacked a single, stratified climate sequence for the Ice Age that extended back further than the Brunhes–Matuyama reversal.

A Cambridge University geophysicist, Nick Shackleton, produced what geologist John Imbrie calls 'the Rosetta Stone' of Ice Age chronology, going back to about three quarters of a million years ago. He analysed a long, deep core from the

The 'Rosetta Stone' of Ice Age climate. A diagram of Ice Age climate change from deep-sea core V28-238 from the southwestern Pacific, which offers a classic view of the climatic fluctuations of the Ice Age revealed by changes in oxygen–isotope ratios. The magnetic reversal here placed at 700,000 years ago, is now dated to 780,000 years before present.

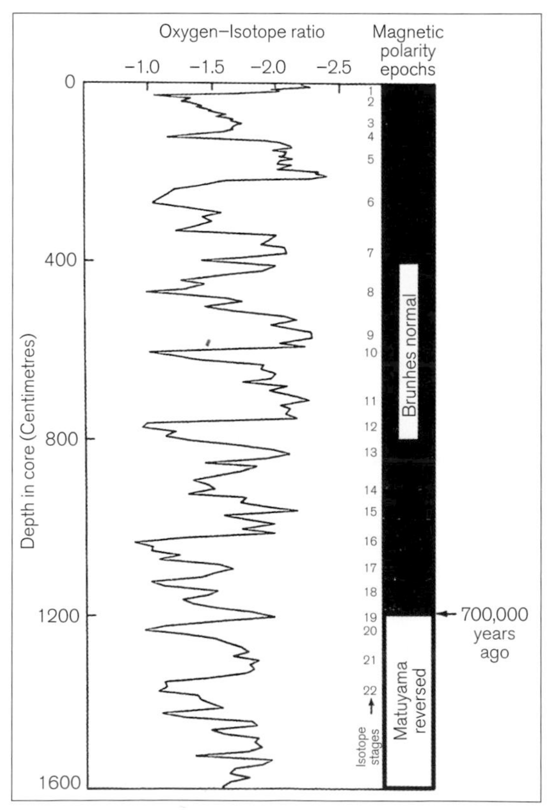

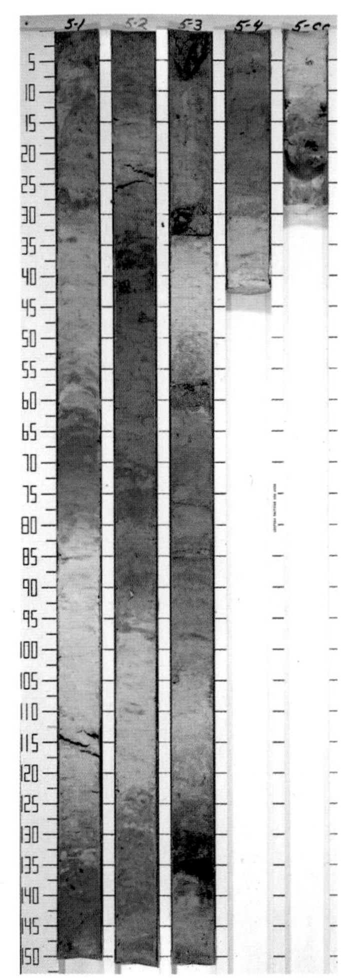

A deep-sea core sample obtained by the *Glomar Challenger* expedition's Deep Sea Drilling Project. The ship retrieved core samples in cores 9 m (30 ft) long, with a diameter of 6.35 cm (2.5 inches). After splitting the core in half lengthwise, one half was archived and the other is still used as a source to answer specimen requests.

shallow western equatorial Pacific known as V28-238, which yielded sediment and molluscs going back to the Brunhes–Matuyama Event. Shackleton developed a mass spectrograph capable, for the first time, of measuring the isotopic variations in the shells of tiny organisms that lived on the sea bed. The resulting curves from V28-238 identified 19 isotopic stages during the Ice Age, dated by radiocarbon dates at the top and by a magnetic reversal at the bottom. Isotope Stage 19 occurred at the Brunhes-Matuyama boundary.

The Shackleton research stimulated new investigations of the Milankovitch curve and Ice Age climatic cycles. Using a statistical method known as spectral analysis, and the CLIMAP time scale, Imbrie and others established the existence not only of a dominant 100,000-year cycle, but of lesser ones about 40,000 and 20,000 years long. After extensive statistical testing using calculations of the earth's eccentricity cycle, they found that climatic oscillations occurred on a 100,000-year cycle corresponding to variations in eccentricity, a 41,000-year cycle resulting from variations in axial tilt, and 23,000- and 19,000-year cycles due to variations in the earth's precession. These dates corresponded very closely with those from isotopic and molluscan studies and also strongly suggested that Milankovitch was correct after all.

These researches, published in 1976, confirmed the essential validity of the astronomical theory first put forward by Adhémar and Croll and later refined by Milankovitch – that variations in the motion of the earth around the sun triggered the multiple glaciations of the Ice Age. However, the mysteries of how the triggering mechanisms worked are still unsolved.

Glomar Challenger, drillship of the Deep Sea Drilling Project, built in 1968 and taken out of service in 1983.

3

How the Age of Ice Began

FIFTY MILLION YEARS AGO the earth was very different from the planet we live on today – it was both warmer and wetter, with rainforest extending all the way up to northern Canada and down to Patagonia. So how did we go from this lush, vibrant earth to the ice-locked cool planet we have now? What caused the beginning of the Ice Age? If you look at a map of the world 50 million years ago and a map of the world today they seem to be the same, but look closer and small differences become apparent. Movements of the continents around the face of the planet are very slow, but minor changes have had a profound effect on global climate.

previous pages Aerial view of a glacier near Juneau, Alaska. This is just one of 38 glaciers that are fed by North America's fifth largest icefield.

What makes a cold planet?

To create an ice age the first thing you need is continents at the poles. Geologists have run simple climate models to demonstrate this idea, showing that if you put all the continents around the Equator – the so-called 'tropical ring world' – the temperature gradient between the poles and the Equator is about 30°C (54°F). This is due to a trick of both the atmosphere and the oceans. The fundamental rule of climate is that hot air rises and cold air drops; this is why in the tropics the land heats up and the air rises, resulting in towering cloud formations developing as the moisture in the air cools and condenses. At the poles it is cold so the opposite happens – the air falls, pushing outwards away from the pole as it hits the ground. So although ice forms at the pole when the sea water freezes, this ice is blown away from the pole towards warmerwater where it melts. This maintains the balance and prevents the temperature of the pole falling below 0°C (32°F). If, however, there is land on the pole or even around the pole, ice can form permanently, and the Equator–pole temperature gradient is much greater – over 65°C (117°F). This is exactly what we have today in the Southern Hemisphere. In contrast, if you consider the Northern Hemisphere, the continents are not actually positioned over the pole, but surround it, so instead of a single huge ice sheet as we have on Antarctica, there is a smaller one on Greenland, and the continents act like a fence keeping all the sea ice in the Arctic Ocean. The Equator–pole temperature gradient of the Northern Hemisphere, therefore, is somewhere between the extremes of an ice-locked Antarctica and a land-free pole: about 50°C (90°F).

The size of the Equator–pole temperature gradient is important for our climate because it is the main driver of the oceanic and atmospheric circulation moving heat from the Equator to the poles. So this temperature gradient defines what sort of climate the world will have. A relatively cold earth such as the one we live in today has an extreme Equator–pole temperature gradient and thus a very dynamic climate. This is why we have hurricanes and winter storms – the climate system is trying to pump heat away from the hot tropics towards the cold poles.

below We live in an age of ice, with large ice sheets on Greenland and Antarctica. In the Arctic, sea ice expands and contracts seasonally and even in summer (as shown below) covers much of the Arctic Ocean.

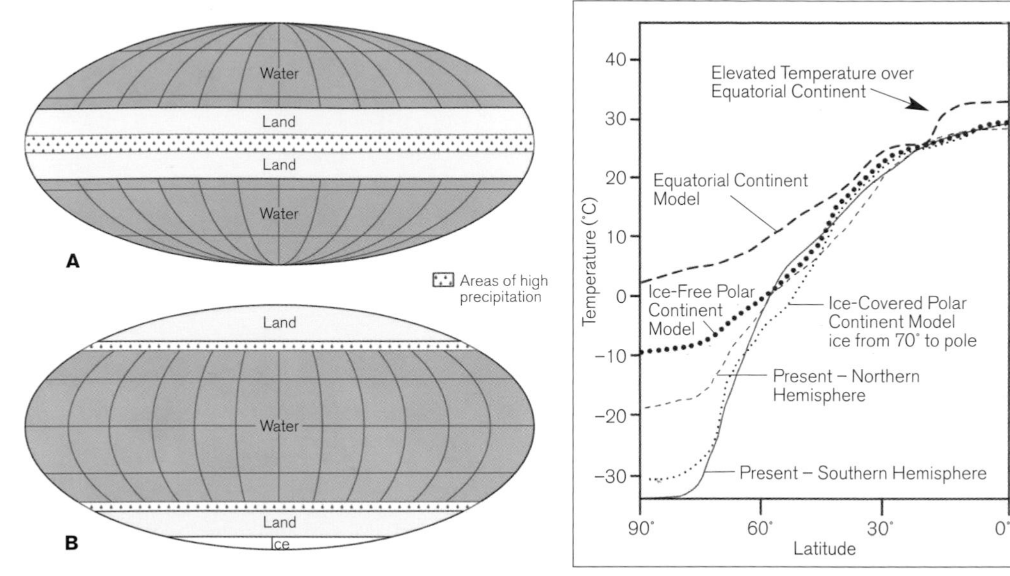

far left The temperature gradient between the Equator and the poles is controlled by where the continents are positioned. If all the continents are at the Equator (A) the gradient is 30°C; if they are at the poles (B) then this gradient can increase to over 60°C.

left The modern-day temperature gradients are shown on the graph, compared with model simulations using the continental configurations in A and B.

For the last 100 million years Antarctica has been positioned over the South Pole and the American and Eurasian continents have surrounded the North Pole. But the Ice Age only began 2.5 million years ago. Therefore there must be other factors controlling the temperature of the earth.

The second thing you need in order to generate an ice age is a means of cooling down the continents on or surrounding the pole. In the case of Antarctica, ice did not start building up until about 35 million years ago; prior to that Antarctica was covered by lush temperate forest – even bones of dinosaurs have been found there. What occurred 35 million years ago was a culmination of minor tectonic movements that caused Antarctica to cool down. South America and Australia are slowly moving away from Antarctica. About 35 million years ago the ocean opened up between Tasmania and Antarctica. This was followed about 30 million years ago by the opening of the Drake Passage, a fearsome stretch of ocean between South

The left of the diagram shows the oxygen isotope record of the deep ocean, which indicates the changes in deep ocean temperature and global ice volume. It shows that for the last 50 million years the world has been getting colder and ice sheets have grown on Antarctica and Greenland. On the right we can compare this record with key climate, tectonic and biological events over the past 70 million years.

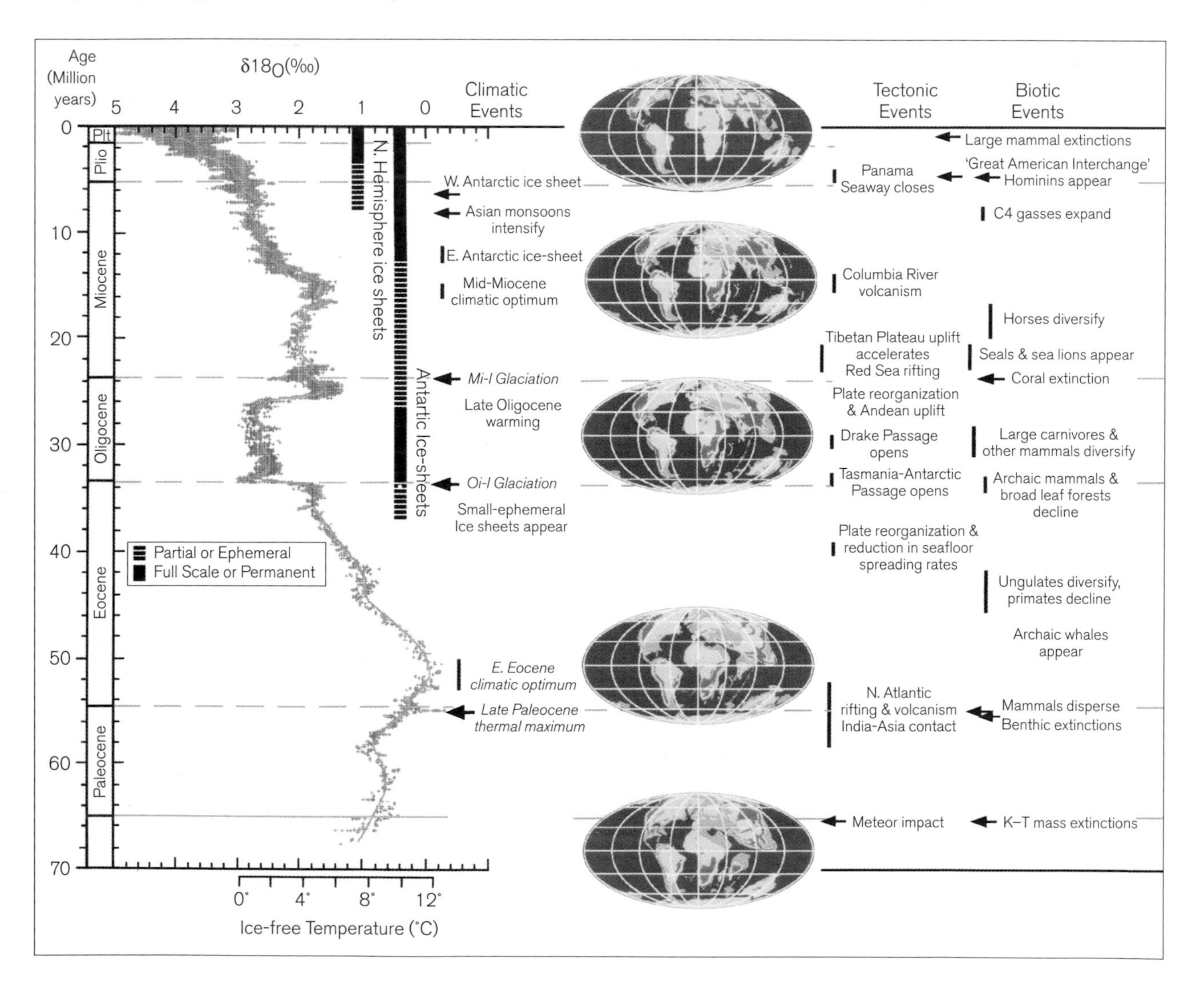

America and Antarctica. This allowed the Southern Ocean to start circulating around Antarctica. The Southern Ocean acts very much like the fluid in a refrigerator: it takes heat from Antarctica as it flows around the continent and releases this heat into the Atlantic, Indian and Pacific Oceans into which it mixes. Thus the opening up of these seemingly small passageways between the continents produced an ocean that can circulate completely around Antarctica, continually sucking out heat from the continent and keeping it cold. So efficient is this process that there is now so much ice on Antarctica that if all of it melted, global sea levels would rise over 70 m (230 ft) – enough to cover the head of the Statue of Liberty. This tectonic cause of the glaciation of Antarctica is also the reason that scientists are confident that the Eastern Antarctic ice sheet, containing about 65 m (213 ft) of sea-level rise, will not melt due to global warming. The same cannot be said of the unstable Western Antarctic ice sheet (see Chapter 8).

The ice-locked Antarctica of 30 million years ago did not last long, however (we do not know why), and 25 million years ago Antarctica ceased to be completely covered with ice and remained that way for the next 15 million years. The question is: why did the world start to cool all over again 10 million years ago, and why did ice start building up in the Northern Hemisphere as well? Palaeoclimatologists believe that the answer lies in the atmosphere, and that relatively low levels of atmospheric carbon dioxide are essential if you are to maintain a cold planet. Computer models have shown that if you have high levels of atmospheric carbon dioxide you cannot get ice to form on Antarctica even with the ocean heat extractor. So what caused carbon dioxide levels to lower, allowing the ice to start growing in the North?

What caused the big freeze?

In 1988 Professor Bill Ruddiman of the University of Virginia and his then graduate student Maureen Raymo wrote an extremely influential paper. They suggested that global cooling and the build-up of ice sheets in the Northern Hemisphere were caused by uplift of the Tibetan-Himalayan and Sierran-Coloradan regions. These huge plateaus, they argued, may have altered atmospheric circulation so that the Northern Hemisphere cooled slightly, allowing snow and ice to build up. However, what they did not realize at the time was that most of the Himalayan uplift occurred much earlier, between 20 and 17 million years ago, and was thus too early to have been the direct cause of the build-up of ice in the North. Following this revelation Maureen Raymo came up with the new, startling suggestion that this uplift in fact contributed to the cooling of the global climate through a massive increase in erosion, which used up carbon dioxide from the air. This, she suggested, occurred because when you create a mountain range you also produce a rain shadow, since moisture in the air condenses as the air is forced up and over the mountain, falling as rain. She argued that this extra rainwater combines with carbon dioxide from the atmosphere to form a weak carbonic acid solution, which dissolves rocks. You can see this as decolouration on any building made of limestone. But interestingly only the weathering of silicate minerals causes a reduction in atmospheric carbon dioxide levels; weathering of carbonate rocks by carbonic acid actually returns carbon dioxide to the atmosphere. Since much of the Himalayas is made up of silicate rocks, there was a lot of rock that could lock up atmospheric carbon dioxide. The new minerals dissolved in the rainwater are then washed into the oceans, and marine plankton use the calcium carbonate to make shells. The calcite skeletal remains of the marine biota are ultimately deposited as deep-sea sediments and are hence lost from the global carbon cycle for the duration of the life cycle of the oceanic crust on which they have been deposited. It's a fast-track way of getting atmospheric carbon dioxide out of the atmosphere and dumping it at the bottom of the ocean. Geological evidence for long-term changes in atmospheric carbon dioxide does support the idea that levels have dropped significantly over the last 20 million years. The only problem scientists have with this theory is answering the question of what stops it. With the amount of rock in Tibet that has been eroded over the last 20 million years, all the carbon dioxide in the atmosphere should have been stripped out. So clearly there are other natural mechanisms that help to maintain the balance of carbon dioxide in the atmosphere.

As atmospheric carbon dioxide lowered between 10 and 5 million years ago, the Greenland ice sheet started to build up. Interestingly, it built up from the south first, due to the warmer water surrounding it. This is because the third thing you must have to cool a planet is a moisture source to build up ice – even in the coldest place in the world there can be no ice if there is no moisture. By 5 million years ago there were huge ice sheets on Antarctica and Greenland very much like today. The Ice Age, when huge ice sheets waxed and waned on North America and northern Europe, did not start until 2.5 million years ago. There is, however, intriguing evidence to suggest that around 6 million years ago these big ice sheets did start to grow. Rock fragments from the continents, eroded by ice and then dumped at sea by icebergs, have been found in the North Atlantic, North Pacific and Norwegian Sea dating to this time. This is the failed attempt at the Ice Age and is all because of the Mediterranean Sea.

The great salt crisis

About 6 million years ago gradual tectonic changes resulted in the closure of the Strait of Gibraltar, leading to the temporary isolation of the Mediterranean Sea from the Atlantic Ocean. During this isolation the Mediterranean Sea dried out several

Satellite view of the Strait of Gibraltar and currents of water rushing through it into the Mediterranean Sea. Five-and-a-half million years ago this Strait closed and the Mediterranean Sea was isolated and started to dry out, producing huge deposits of salt. It was not until 200,000 years later that the Strait re-opened and the Atlantic Ocean flooded in.

times, depositing vast salt (evaporite) deposits up to 3 km (1.9 miles) thick. Just imagine this huge version of the Dead Sea where just a few metres of seawater covers a vast area. This event, during which nearly 6 per cent of all dissolved salts in the world's oceans were removed, is called the Messinian Salinity Crisis (named after Messina in Sicily) and generated a profound change in the global climate. By 5.5 million years ago the Mediterranean Sea was completely isolated and was a salt desert. This was roughly the same time as the world was trying to enter the Ice Age. But at about 5.3 million years ago the Strait of Gibraltar reopened and caused what is known as the Terminal Messinian Flood. The gigantic waterfall at the Strait of Gibraltar as Atlantic Ocean water poured into the Mediterranean would have dwarfed anything we see on earth today. It also meant vast quantities of salt were dissolved and pumped back into the world's oceans via the Mediterranean-Atlantic gateway, affecting the way the oceans circulated and stopping the Ice Age in its tracks. This demonstrates the crucial role the circulation of the oceans play in determining the climate of the planet.

The global ocean conveyor belt

Salt and heat are the main controls on the deep circulation of the ocean. Today the tropical sun heats surface water in the Caribbean, causing there to be a lot of evaporation and thus starting the hydrological cycle – the flow of water around the

planet. All this evaporation leaves the surface water enriched in salt. The hot, salty surface water is pushed by the winds out of the Caribbean along the coast of Florida and into the North Atlantic Ocean. This is the start of the famous Gulf Stream. The Gulf Stream is about 500 times the size of the Amazon River at its largest and flows along the coast of the USA and then across the North Atlantic Ocean past the coast of Ireland, past Iceland and up into the Nordic Seas. As the Gulf Stream flows northwards it cools down. The combination of a high salt content and lower temperature makes the surface water heavier or denser. Hence, when it reaches the relatively fresh (i.e. less salty) oceans north of Iceland the surface water has cooled sufficiently to become dense enough to sink into the deep ocean, where it forms the North Atlantic Deep Water, a water mass that flows down the Atlantic Ocean, joins other deep water formed around Antarctica and then flows into the Indian and Pacific Oceans where it is eventually forced to the surface and returns as surface water to the Caribbean. For a parcel of water to flow around the whole of this great global ocean conveyor belt takes about 1,000 years, and is like the slow heartbeat of global climate.

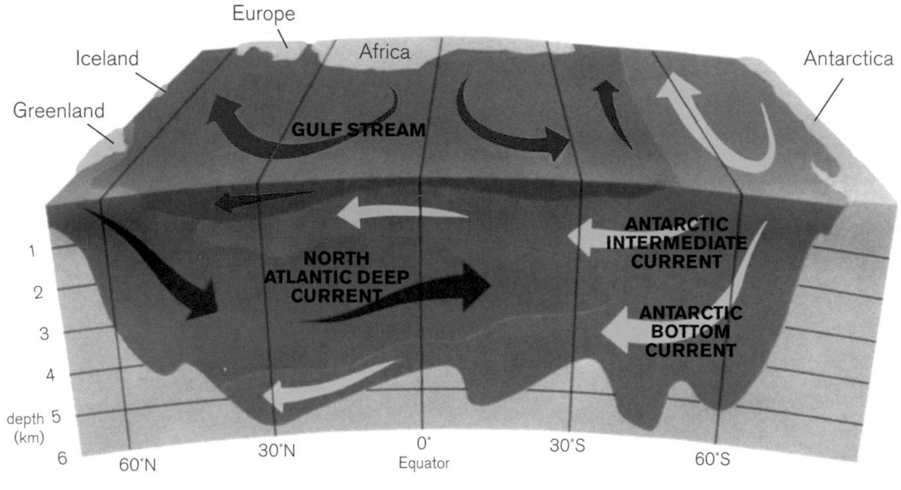

above There are two sources of deep water in the ocean: North Atlantic Deep Water (NADW), formed in the north, and Antarctic Bottom Water (AABW), formed in the south. The balance between these two water masses controls global climate.

The 'pull' exerted by the sinking of this dense North Atlantic Deep Water helps maintain the strength of the warm Gulf Stream, ensuring a current of warm tropical water flowing into the northeast Atlantic, sending mild air masses across to the European continent. It has been calculated that the Gulf Stream delivers 27,000 times the energy of all of Britain's power stations put together. If you are in any doubt about how beneficial the Gulf Stream is for the European climate, compare the winters at the same latitude on either side of the Atlantic Ocean, for example, London with Labrador or Lisbon with New York. An even better comparison is between Western Europe and the West Coast of North America, which have a similar geographical relationship between the ocean and continent; so think of Alaska and Scotland, which are at about the same latitude.

The heat released by the Gulf Stream not only keeps Europe warm but is also sufficient to keep the planet from cooling down. As we will find out in the section below, 5 million years ago the Gulf Stream and deep ocean circulation were not as strong as they are today. This is because fresher Pacific Ocean water was still able to leak through the Panama ocean gateway, since the separate continents of North and South America had not yet fully come together. This weak Gulf Stream meant that the climate was much cooler, but the sudden massive increase in salt due to the Terminal Messinian Flood increased the salinity of the North Atlantic Ocean, ensuring a very vigorous Gulf Stream and sinking water in the Nordic Seas. With all this tropical heat being efficiently pumped northwards, the slide into the Ice Age was halted. It was another 2.5 million years before global climate was ready to try again.

opposite top The world's oceans are linked into one large circulation called the ocean conveyor belt. Warm salty water cools in the North Atlantic Ocean and sinks, forming deep water (blue on map), which can circulate down the Atlantic Ocean, around Antarctica and ends up rising to the surface in either the Indian or North Pacific oceans. The surface water (red on map) then returns to the North Atlantic Ocean ready to sink again.

left The passing of the warm waters of the Gulf Stream so close to the British Isles keeps the winters relatively mild. This allows plants to grow which are usually found much further south – an example of this is Inverewe gardens in the North of Scotland, which lie at 57.8 degrees north, about the same latitude as Hudson Bay, Canada.

The Panama paradox

Another important tectonic control that geologists believe to be a trigger for the Ice Age is the closure of the Pacific-Caribbean gateway. Two leading German scientists, professors Gerald Haug and Ralf Tiedemann, have used evidence from ocean sediments to suggest that the Panama ocean gateway began to close at 4.5 million years ago and finally closed about 2 million years ago. The closure of the Panama gateway, however, causes a paradox as it would have both helped and hindered the start of the Ice Age. First, the reduced inflow of Pacific surface water to the Caribbean would have increased the salinity of the Caribbean, as we have seen, because Pacific Ocean water is fresher than its counterpart in the North Atlantic Ocean. This would have increased the salinity of water carried northwards by the Gulf Stream and North Atlantic Current, and would have enhanced deep-water formation and heat transport to the high latitudes, thus hindering ice sheet formation. Yet secondly the enhanced Gulf Stream would have pumped a lot more moisture northwards, stimulating the formation of ice sheets. Thus the Ice Age could start at a warmer temperature because of all that moisture being pumped northwards ready to fall as snow and to build up ice sheets.

Why 2.5 million years ago?

Tectonic forcing alone cannot explain the amazingly fast onset of the Ice Age. Work undertaken by Mark Maslin on ocean sediments suggests there were three main steps in the transition. (The evidence is based on rock fragments ripped off the continent by ice and deposited in the adjacent ocean basin by icebergs.) First, ice sheets started growing in the Eurasian Arctic and northeast Asia regions at approximately 2.74 million years ago, with some evidence of growth of the northeast American ice sheet. Second, an ice sheet began to build up on Alaska at 2.7 million years ago, and third, at 2.54 million years ago the northeast American ice sheet – the biggest of them all – reached its maximum size. So in less than 200,000 years the planet went from the warm, balmy conditions of the early Pliocene, the 'golden age of climate', to the Ice Age.

Therefore the timing of the start of the Ice Age must have a cause other than tectonic plate movements. It has been suggested that changes in orbital forcing, i.e. the way the earth spins round the sun, may have been an important mechanism contributing to the global cooling. The details of the earth's numerous wobbles, and how they cause the waxing and waning of individual glaciations, are discussed in Chapter 4. But though these wobbles occur on the scale of tens of thousands of years, there are much longer variations as well. One of the most important is obliquity, or tilt, which is the wobble of the earth's axis of rotation up and down. Or, put another way, it is the tilt of the earth's axis of rotation with respect to the plane of its orbit.

Over a period of 41,000 years the earth's axis of rotation will lean a little bit more towards the sun and then a little less. It's not a large change, varying between 21.8° and 24.4°. It is the tilt of the axis of rotation that gives us the seasons, since in summer the hemisphere that is tilted towards the sun is warmer because it receives more than 12 hours of sunlight, and the sun is higher in the sky. At the same time, in the opposite hemisphere, axis of rotation is tilted away from the sun and so that region is plunged into winter, when it is colder and receives less than 12 hours of sunlight, and the sun is lower in the sky. Hence the larger the obliquity, the larger the difference between summer and winter. Over a period of 1.25 million years the amplitude of the tilt changes. Both times the earth tried to enter the Ice Age at 5 million years and 2.5 million years ago the variation of tilt was at its largest value. This made the changes in each season very large. Especially important is having cold summers in the North so that ice can survive the summer and develop into ice sheets.

The tropics react to the Ice Age

The onset of the Ice Age did not just affect the high latitudes. It seems that half a million years after the start of the Ice Age things also changed in the tropics. Before 2 million years ago there appears to have been very little East–West sea surface temperature gradient in the Pacific Ocean, but there was afterwards. This shows a switch in the tropics and subtropics to a modern mode of circulation, with relatively strong atmospheric circulation and cool subtropical temperatures. The atmospheric circulation that occurs between the East and the West is known as Walker circulation, and it leads to one of the most important and mysterious features of global climate: the periodic switching of the direction and intensity of ocean currents and winds in the Pacific. Originally known as El Niño ('Christ child' in Spanish), as it usually appears at Christmas, and now more normally known as ENSO (El Niño–Southern Oscillation), this phenomenon typically occurs every three to seven years, and may last from several months to more than a year. ENSO is in fact an oscillation between three climates, the 'normal' conditions, La Niña, and El Niño. The 1997–98 El Niño conditions were the strongest on record and caused droughts in southern USA, East Africa, northern India, northeast Brazil and Australia. In Indonesia, forest fires burned out of control in the very dry conditions. In California, parts of South America, Sri Lanka, and east-central Africa there were torrential rains and terrible floods. El Niño conditions have been linked to changes in the monsoon and to droughts all over the world, as well as the position and occurrence of hurricanes in the Atlantic Ocean. For example, it is thought that the poor prediction of where Hurricane Mitch made landfall in 1998 was because the ENSO conditions were not considered and the strong trade winds helped drag the storm south across Central

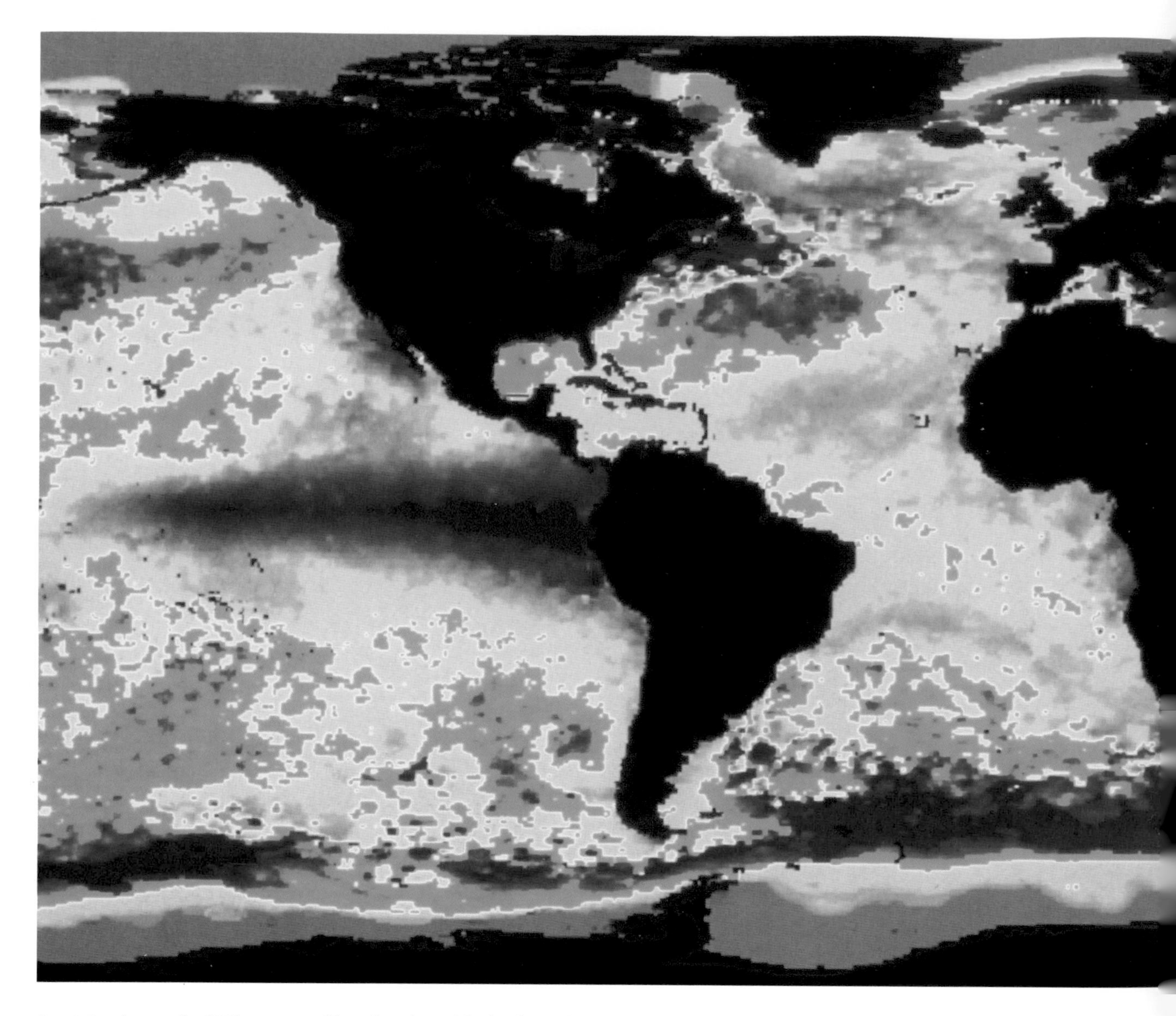

America instead of West as predicted. It is unlikely that El Niño would have existed before 2 million years ago because there was a relatively weak Walker circulation. So the Ice Age influenced both tropical and subtropical climate, and may even have given birth to the devastating El Niño.

Conclusion

We still do not know precisely what caused the cooling-down of earth over the last 50 million years. However, lots of very small changes in tectonics seem to have conspired to push the planet to colder and colder conditions. We know that having continents at or around the poles is essential to create a cold planet. Scientists also think it is plausible that the Tibetan uplift caused long-term cooling due to changing

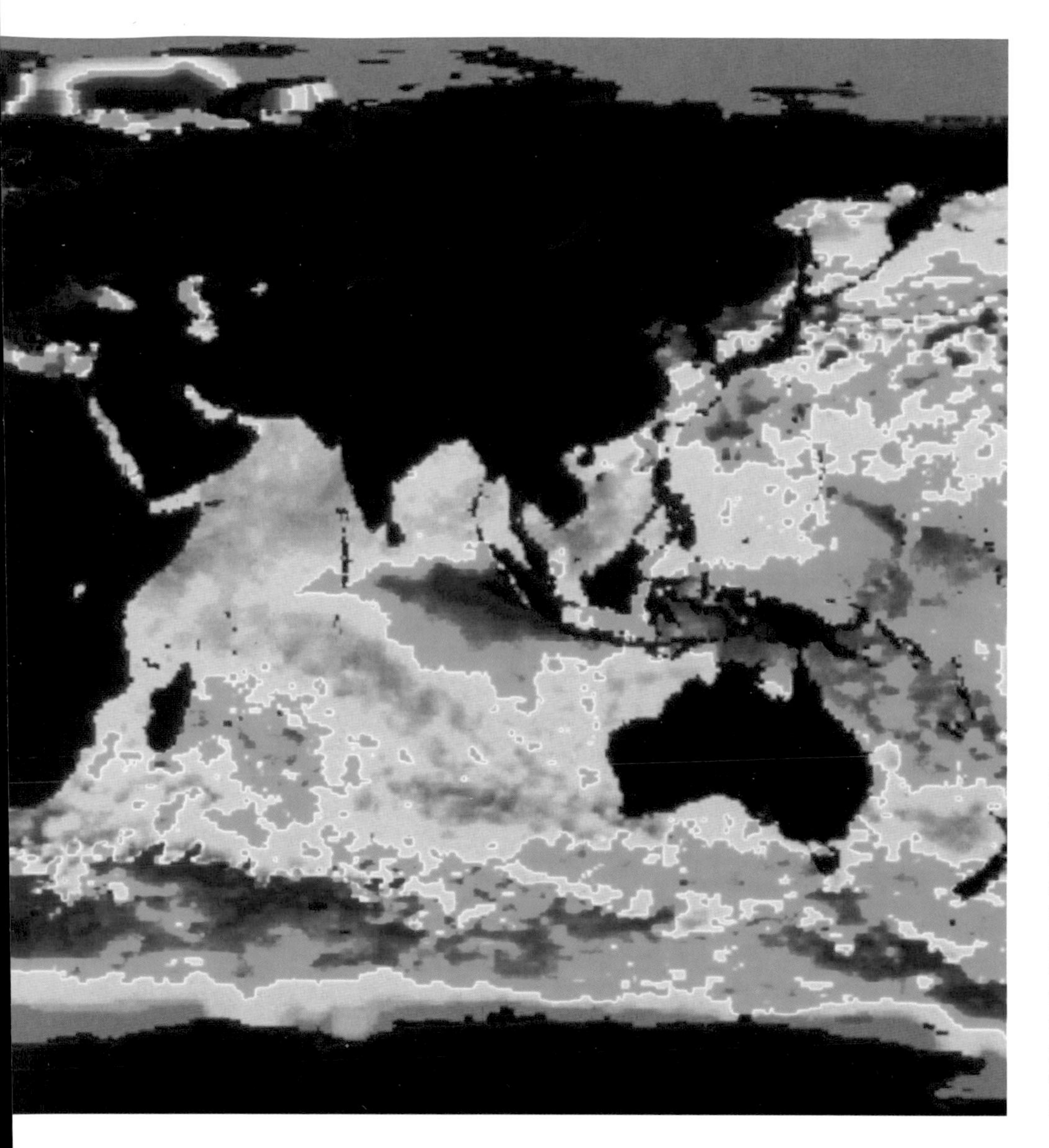

El Niño is a global climate event that occurs every 3 to 7 years. The changes in ocean temperature are marked and shown here in a satellite picture. El Niño causes droughts in Amazonia, floods in California, forest fires in Indonesia and changes the paths of hurricanes, but scientists believe that the phenomenon may not have existed before about 2 million years ago.

atmospheric circulation and lowered atmospheric carbon dioxide levels. But these factors alone were not enough to cause an ice age. As the 'failed attempt' at the Ice Age shows us, ocean currents have an enormously strong effect on global climate. The drying-out and subsequent flooding of the Mediterranean Sea, affecting ocean currents, seems to have been instrumental in aborting this first attempt at the Ice Age 5 million years ago. The closure of the Panama ocean gateway may then have delayed the onset of the Ice Age, but ultimately provided the moisture that allowed it to be very intensive when it did arrive. The global climate system seems to have reached a threshold at about 2.5 million years ago. The final piece in the complex climatic jigsaw was put in place when long-term changes in the earth's wobbling orbit pushed global climate across this threshold, producing the phenomenon we call the Ice Age.

4
The Climatic Rollercoaster

For the last 2.5 million years the climate of the earth has been dominated by the coming and going of the great ice sheets. Only 21,000 years ago these ice sheets were so thick that there was 3.2 km (2 miles) of ice piled up on both North America and northern Europe; we only have to imagine Chicago or Edinburgh under such a huge weight of ice to see how dramatically the world of today differs from that of the last glacial period. And it was not just the extreme northern and southern areas of the globe that were affected: during glacial periods average global temperatures were 6°C (11°F) lower than today and global sea levels were 120 m (394 ft) lower; the total weight of all the plants on the land was reduced by as much as half, and atmospheric carbon dioxide was reduced by a third and

previous pages A modern-day iceberg off the coast of Greenland gives us a glimpse of how the North Atlantic Ocean would have looked 18,000 years ago, when ice sheets 3.2 km (2 miles) thick sat on North America and Europe, sending out huge armadas of icebergs into the sea.

atmospheric methane by a half. These periods of extreme cold also dramatically changed the landscape of the planet, which was shaped by the erosion and deposition of sediment by the huge ice sheets. Major rivers were re-routed, mountains were cut in half and land bridges appeared as the oceans dropped, connecting continents and allowing species to colonize new lands. In this chapter we explore what the earth was like during these times of momentous climatic and environmental change and look at what caused the waxing and waning of the huge ice sheets.

above Marine sediments, such as the core being extracted here, are essential in our investigation of the Ice Age. As marine sediment accumulates slowly over thousands of years it creates an archive of the changes in past climate.

Investigating the Ice Age

To recap what we have learnt in the previous two chapters: over the years scientists have used numerous methods of investigation to form theories about the Ice Age, and as techniques have progressed, so our knowledge of what the Ice Age climate was like has grown ever more detailed. From James Ussher's attribution in 1658 of the features of the land to Noah's Great Flood, to Horace-Bénédict de Saussure's reasoning in 1787 that erratic boulders of Alpine rocks had been moved down the slopes of the Jura Range by glaciers (a theory later expanded by Louis Agassiz in 1837), early attempts to explain the mysterious features of the world's glacial landscapes gave little indication of the real dynamics of the world's climate during glacial times, nor of the timings of events. In 1909 Penck and Brückner, looking at glacial deposits, edged closer to a detailed understanding of the dynamics of the Ice Age with their theory that there were four major glacial periods: the Günz, Mindel, Riss and Würm.

The problem with the early studies of the Ice Age is that terrestrial or land-based evidence is poor because it is discontinuous, and therefore it is not always possible to determine how the many different moraine deposits are related to each other. Also, land sediments are hard to date, so it is difficult to tell when the deposits were laid down. Finally, through erosion ice sheets can destroy signs of previous glaciations, so it is entirely possible that evidence of whole glacial periods may have been wiped out. Hence it was not until the 1960s, when long continuous sediment cores were being recovered from the oceans, that it was realized how many glacial periods there really had been. We now have the ability to drill in oceans over 6 km (4 miles) deep

opposite A glacial moraine from the Cowlitz glacier on the southeast flank of Mount Rainier in Washington State, USA. Many features of the land around us have been shaped by the Ice Age. Here the debris carried by the glacier has built up into great mounds which are left behind as the glacier retreats.

and still recover nearly a kilometre (over half a mile) of sediment from below the sea bed. By studying these marine sediments scientists have documented 50 glacial periods that occurred in the last 2.5 million years. It is now known that between 2.5 million years and 1 million years ago these glacial–interglacial cycles occurred every 41,000 years, and since 1 million years ago they have occurred every 100,000 years.

Another revolution caused by the marine records was the switch away from names to numbers for labelling different climate periods. Until the 1970s nearly every country had its own names for the different interglacial and glacial periods. So complicated was the system that a little cottage industry was dedicated to working out which names were equivalent to which! One of the pioneers of long marine sediment records was the British scientist and climatologist Nick Shackleton, introduced in Chapter 2, who, as we have seen, devised a simple approach to naming the periods. Each interglacial and glacial period is numbered starting from our own interglacial as number 1 and going back in time, so that any climate period with an odd number is an interglacial and any period with an even number is a glacial period. The Holocene, therefore, is 'number 1' and the last glacial maximum about 21,000 years ago is '2'. This simple approach means that scientists have a common language with which to discuss different periods across the whole planet.

Anatomy of the last ice age

We now know that the glacial–interglacial cycles that characterize the Ice Age are driven by changes in the earth's orbit around the sun. The complexity of how changes in the earth's orbit translate into climate change on earth means that the timing of the glacials and interglacials over the Ice Age has not been absolutely regular. The periods have varied both in duration and severity, and over the past 2.5 million years there have been notable changes in the pattern of the cycles.

As we have seen, between 2.5 and 1 million years ago the glacial–interglacial cycles occurred approximately every 41,000 years, and neither the warmth nor the coldness was extreme. About 1 million years ago the glacial–interglacial cycles began to take place over a much longer time period of approximately 100,000 years. This elongation was caused primarily by the increased time the planet spent in glacial conditions. Greater amounts of ice accumulated during each of the glacials. The amplitude of the warm interglacials and cold glacials increased, with both extreme prolonged glacials, and often also extremely warm interglacial periods. Interglacials as warm as the last five can only be found at around 1.1, 1.3 and before 2.2 million years ago. As well as being unusually warm, the five interglacial episodes that occurred after half a million years ago (at about 420,000, 340,000, 240,000, 130,000 and 12,000 years ago) are also associated with much higher levels of carbon dioxide than any that preceded them – about 290–300 parts per million (ppm), compared with about 260 ppm in earlier episodes. The longest ever interglacial started 430,000 years ago and lasted for over 28,000 years. During this time temperatures were warmer than today and global sea levels were up to 15 m (49 ft) higher, suggesting a significant melting of both the Greenland and the Western Antarctic ice sheets. What is interesting to note is that before 1 million years ago the transitions between glacial and interglacial periods were regular and balanced. After 1 million years ago the climate records become 'saw-toothed', with a long 80,000-year period of cooling to a glacial maximum and then less than 4,000 years' rebound to interglacial conditions. The saw-toothed climate cycles are also irregular, with the duration of the cycles varying from 84,000 to 119,000 years in total.

If we focus on just the last glacial period, only 21,000 years before today, we can see what a considerable difference glaciation made to the climate of the planet. In North America there was nearly continuous ice covering the continent from the Pacific to the Atlantic Ocean. It was made up of two separate ice sheets, the Laurentide ice sheet in the east, centred on Hudson Bay, and the Cordilleran ice

The map below shows the maximum extent of the huge ice sheets which grew during the glacial periods. Ice built up on North America, Europe and South America and expanded on Antarctica. As ice accumulated on the continents, sea levels lowered dramatically, allowing humans and animals to colonize previously unreachable parts of the world. One of the most significant connections formed by the Ice Age was the Bering land bridge, which allowed humans to pass from northeast Asia into Alaska and down into North and then South America.

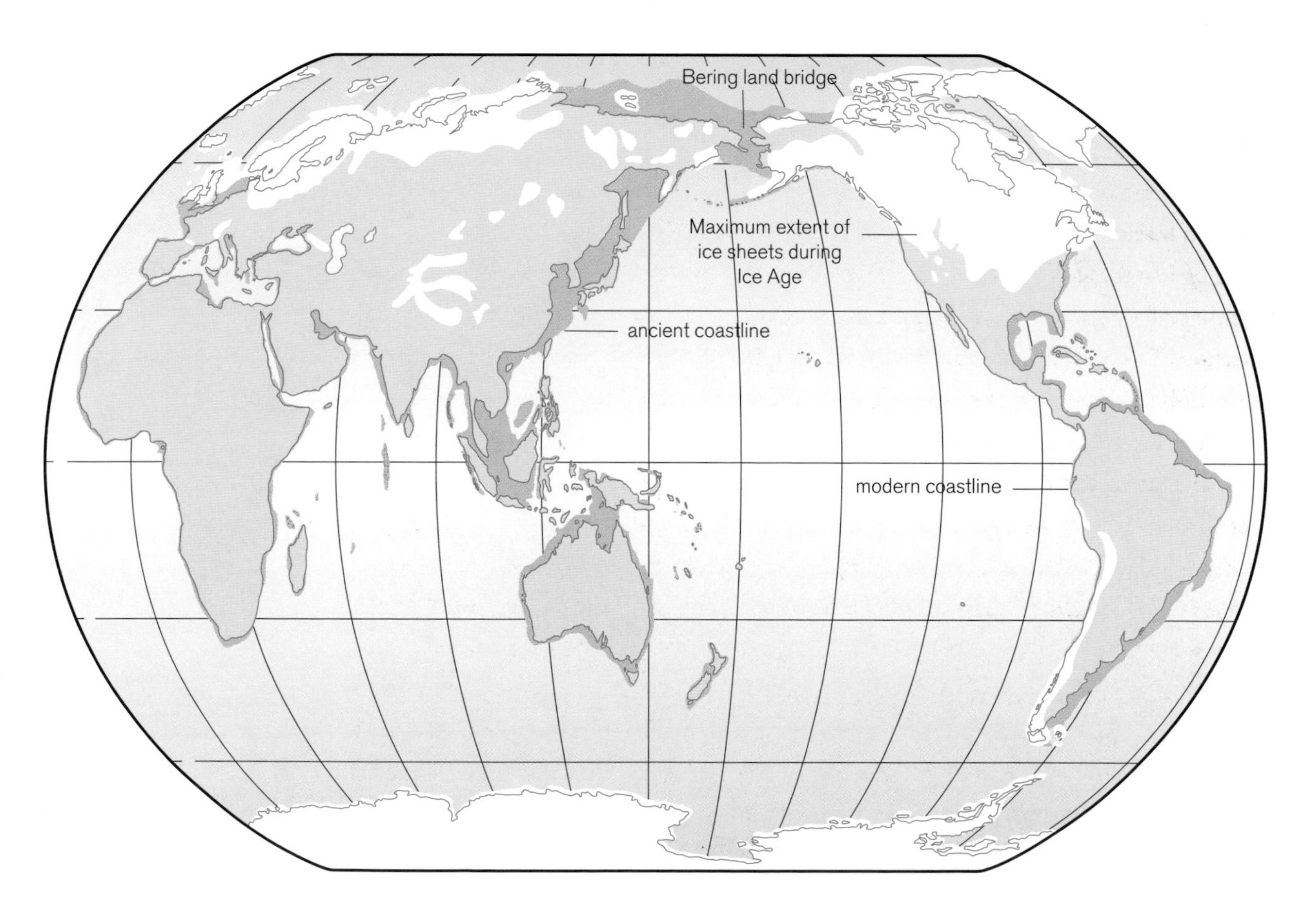

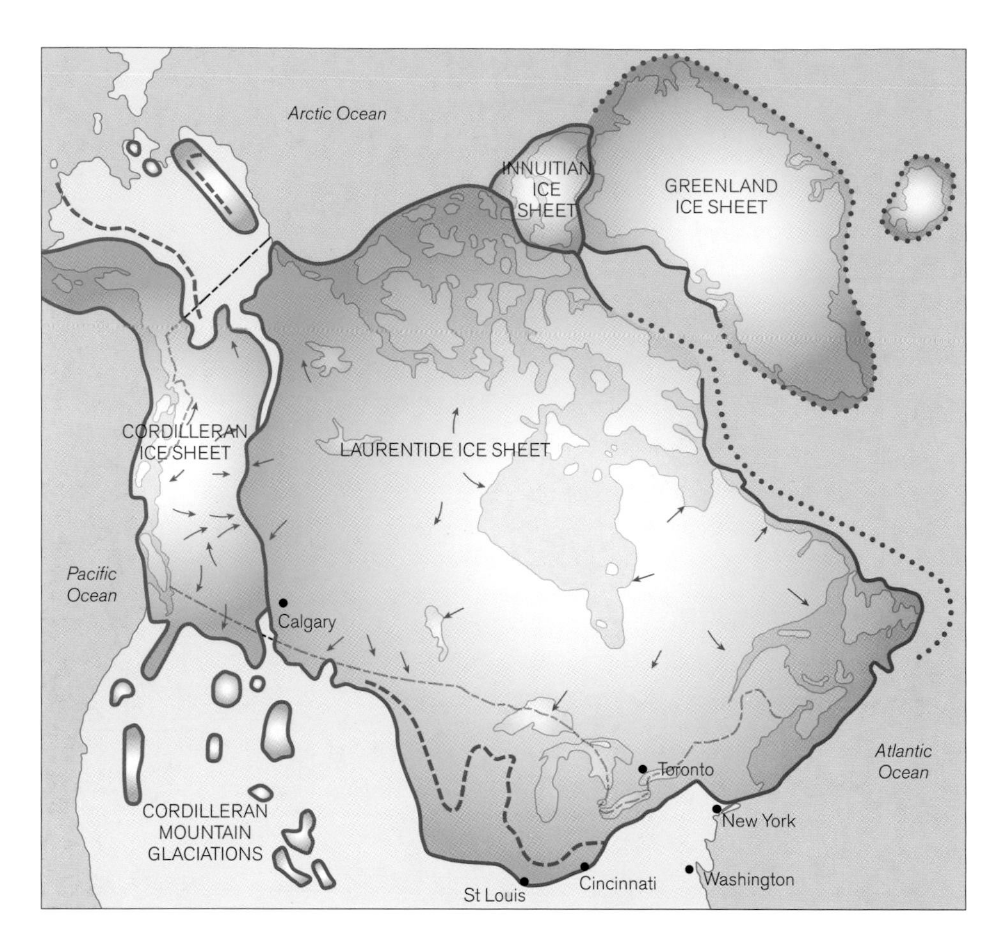

The three maps provide a detailed view of where the ice sheets built up during glaciations on North America, South America and Britain.

below On the map of Britain two limits of the British ice sheet are given. The first, solid line indicates the position of the ice sheet during the last glaciation, which reached its maximum extent about 18,000 years ago. However, during the glaciation before that one, about 140,000 years ago, the ice sheet extended much further south (see dotted line) and reached Bristol and North London.

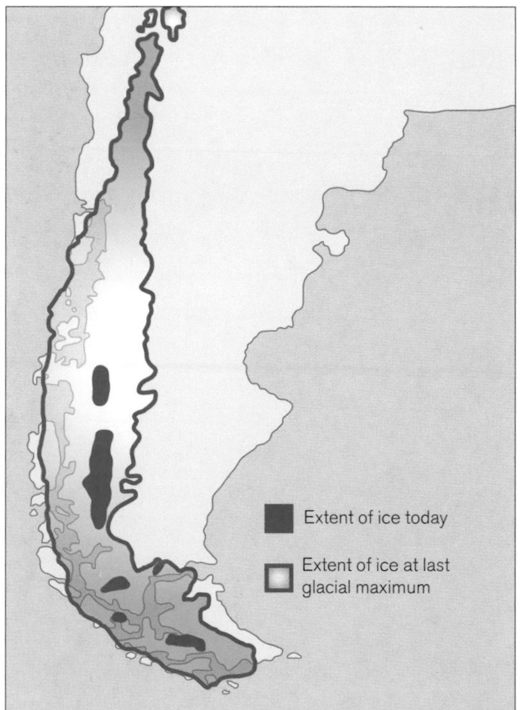

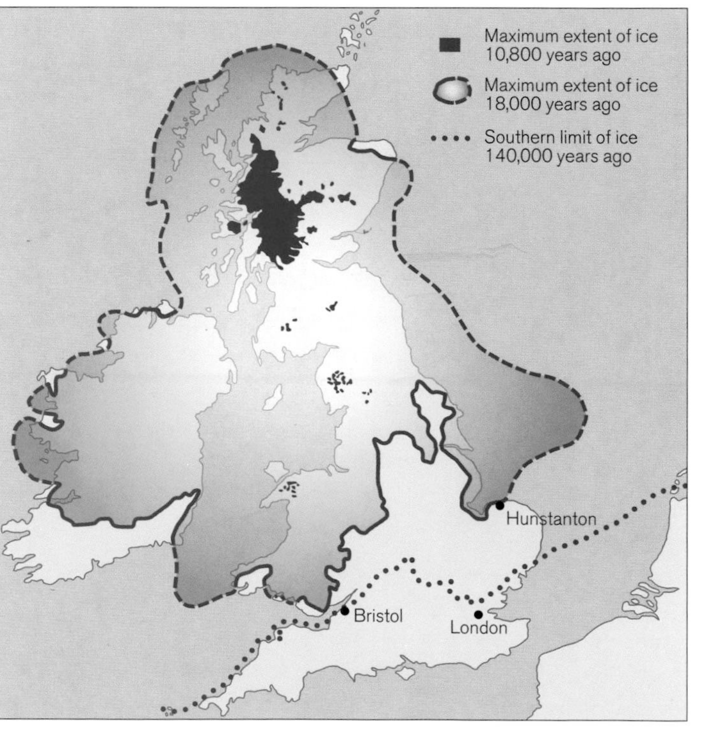

sheet in the west, associated with the coastal ranges and the Rockies. The Laurentide ice sheet covered over 13 million sq. km (5 million sq. miles) of land and reached over 3,300 m (10,827 ft) in thickness at its deepest over the Hudson Bay. At its maximum extent it reached New York, Cincinnati, St Louis and Calgary. In contrast, the smaller Cordilleran ice sheet covered only 2.5 million sq. km (965,000 sq. miles) of land and reached a maximum thickness of 2,400 m (7,874 ft). Greenland was also 30 per cent bigger than today during the last glacial period and was linked to the top of the Laurentide ice sheet by the small Innuitian ice sheet.

In Europe there were two major ice sheets, the Scandinavian and the British ice sheets, and a minor one covering the Alps. On average the British ice sheet covered about 340,000 sq. km (130,000 sq. miles) during each glacial period, and during many glacials it merged with the Scandinavian ice sheet. It had key local centres of ice over the Scottish Highlands, the Lake District, the southern uplands, the Pennines, and the Welsh and Irish mountains, and during the last glacial period it reached halfway down the British Isles to just past Hunstanton in Norfolk. In earlier glacial periods the ice sheet came all the way down to Bristol and North London. The Scandinavian ice sheet was much larger than the British ice sheet, covering an area of 6.6 million sq. km (2.5 million sq. miles) and extending all the way from Norway to the Ural Mountains in Russia. There is evidence that this ice sheet may even have joined with the ice over Spitzbergen, north of the Arctic Circle. Although there is still controversy about the extent of ice in Siberia and northeast Asia, estimates suggest that at least 3.9 million sq. km (1.5 million sq. miles) was covered with ice, over ten times the area covered by the British ice sheet. In the Alps the glaciers extended down to 500 m (1,640 ft) above sea level in the north and 100 m (328 ft) above sea level in the south during the last glacial period, and may have reached a maximum thickness of 1,500 m (4,921 ft).

It is important not to forget the Southern Hemisphere, since there were also significant ice sheets on Patagonia, South Africa, southern Australia and New Zealand. In addition to this, the Antarctic ice sheet expanded by about 10 per cent and seasonal sea ice extended an additional 800 km (500 miles) away from the continent. By comparing the distributions of these ice sheets we can see the marked difference between an interglacial like the present and the last glacial period. Today 86 per cent of the continental ice on the planet is on Antarctica, with another 11.5 per cent on Greenland. Twenty-one thousand years ago things were very different, with two and a half times as much ice on the land as today; 32 per cent of it on Antarctica, 35 per cent on North America, 15 per cent on Scandinavia, 5 per cent on Greenland, 9 per cent on northeast Asia and 2 per cent on the Andes (diagram p.70). The British ice sheet represented less than 0.7 per cent of the total ice on the planet.

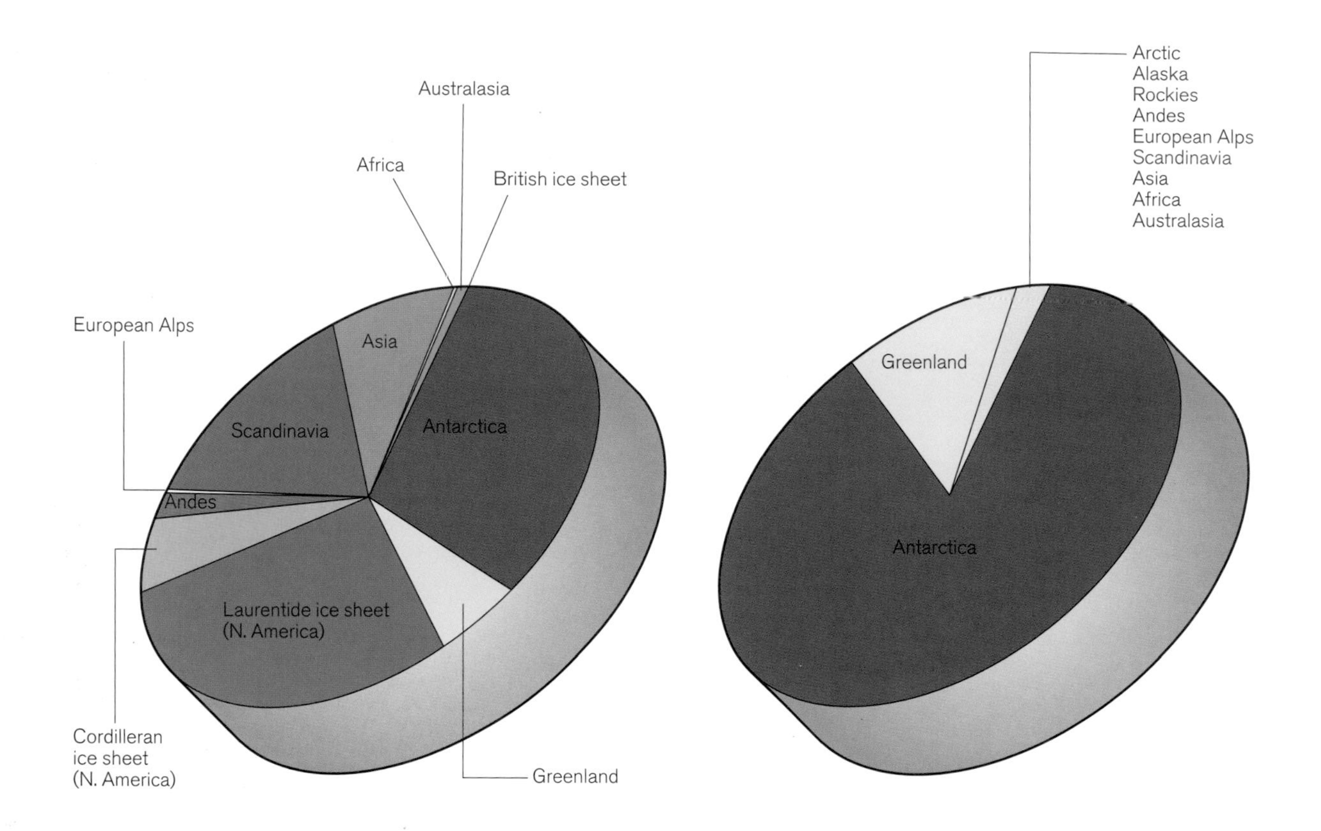

The pie charts above represent the distribution of ice on earth during the last glacial maximum, 21,000 years ago (left) and today (right). During the last glaciation, glaciers probably covered around 45 million sq. km (17 million sq. miles); today they cover only 15.7 million sq. km (6 million sq. miles).

It is difficult to imagine the vast quantity of ice that was locked up in these ice sheets. One way of understanding it is to consider the oceans. The oceans cover over 70 per cent of our planet, and so much water was sucked out of them and locked up in the glacial ice sheets that the sea level dropped by over 120 m (394 ft). This is about the height of the London Eye. If all the ice on Antarctica and Greenland today melted it would raise the sea level by a staggering 85 m (279 ft), enough to drown many of the most densely populated regions on earth.

Ice shapes the land

During the glacial periods the presence of the huge ice sheets profoundly affected the local climate and environment. First the great boreal forests of the high latitudes were devastated as the land they occupied was shrouded under ice. Second, around the margins of each ice sheet developed huge areas of permafrost. This is ground that is permanently frozen with only the top few metres melting every summer. Throughout Britain, northern Europe, northern Asia and North America evidence of permafrost can be found, including ice wedges and patterned ground. These

above This distinctive landscape feature, known as patterned ground, is found in areas with permafrost, where the soil is permanently frozen apart from the very top layer. The summer thawing and winter refreezing of the surface layer cause great cracks to appear, giving rise to these patterns.

permafrost areas were also covered with seasonal tundra-type vegetation, and at the southern edge with permanent steppe-tundra vegetation. However, the areas affected were so much further south than similar areas today that the environment was very different. This is because the sun's energy during the summer period is much more intense at these latitudes than it is further north. The vegetation type that grew in these areas, therefore, has no modern analogue and has its unique name, taiga. It combines steppe and tundra species with conifer and sometimes even broad-leaf trees.

The huge continental ice sheets had an astonishing impact on our landscape. There are very few places in the temperate latitudes that have not been affected by glaciation. Nearly everyone has seen the dramatic effects of ice sheets on the landscape, partly because many of us have travelled through European or North American landscapes, but also because glacial landscapes are often used as dramatic backdrops to movies, such as the *Lord of the Rings* trilogy, filmed in New Zealand. That mountainous, wild, extreme landscape is the result of an ice sheet grinding away over the islands for thousands of years.

overleaf The dramatic landscape of New Zealand shows the classic features of an ice-eroded land. Sharp mountain peaks contrast with the wide U-shaped valleys cut by glaciers.

U-shaped valleys, such as this one in the San Juan Mountains, Colorado (*above left*), are cut by glaciers flowing down from the mountains. When the sea level rises at the end of the glacial period these U-shaped valleys can become flooded, forming fjords, as in Norway (*above right*).

In Britain there are U-shaped valleys where glaciers have carved out a route down from the ice sheet towards the sea. In Scotland, Norway, Greenland, Canada, New Zealand and Patagonia the rising sea level at the end of the Ice Age flooded many of these U-shaped valleys, forming spectacular fjords. Other major features of the landscape, such as the long lines of hills known as 'end moraines' that Louis Agassiz studied, were formed from material dumped by the ice sheets. In Yorkshire and the middle of England there are uniquely shaped hills called drumlins, which formed as the ice pushed up and over soil and rock, squashing them into an egg shape. Even the current position of the river Thames is due to ice. Previously it ran through St Albans to the north of London and met the North Sea in Essex. The last-but-one glaciation was so intense in Europe that the ice sheet made it down to Finchley in North London – there is an area still visible today called the Finchley depression where the ice sheet ground to a halt. This re-routed the river Thames to its current path. So the geography of London was primarily controlled by the Ice Age.

In the USA the paths of many major rivers were altered both by the location of the ice sheets but also by the huge amount of meltwater that burst from them as they melted 12,000 years ago. The pathways of the St Lawrence and Mississippi rivers are relics of the great floods at the end of the last glacial period. The next time you are driving through or flying over England or America, try to work out which features of the landscape were caused by the effects of the Ice Age and which have been caused more recently. A good example of this overlap is a glacier-cut U-shaped valley with a river running through it cutting its own V-shaped valley in the floor of the older valley.

The glacial periods also affected areas much further away from the ice sheets. Global temperatures were 6°C (11°F) lower than today, but this temperature drop was not evenly distributed, the high latitudes cooling by as much as 12°C (22°F), although even the tropics cooled by between 2° and 5°C (4° and 9°F). Glacial periods were also very dry, with huge amounts of dust in the atmosphere (see Chapter 1). In northern China, the eastern USA, central and eastern Europe, central Asia and Patagonia there are deposits of hundreds of metres of dust, called loess deposits, which built up during glacial periods.

The geography of the earth was transformed as the lowering of global sea levels by 120 m (394 ft) meant that the continents changed shape. Islands such as Britain became part of the mainland. Imagine being able to walk across the English Channel to France: the only thing that would stop you would be the huge new river running down the centre of what is now the English Channel, taking water from the Thames, Rhine and Seine out to the Atlantic Ocean. All around the world, land bridges were

Glacial periods are very dry because so much water is locked up in ice sheets. During each glaciation vast quantities of dust are generated, and these form huge deposits all over the world. The map below shows where these loess deposits can be found.

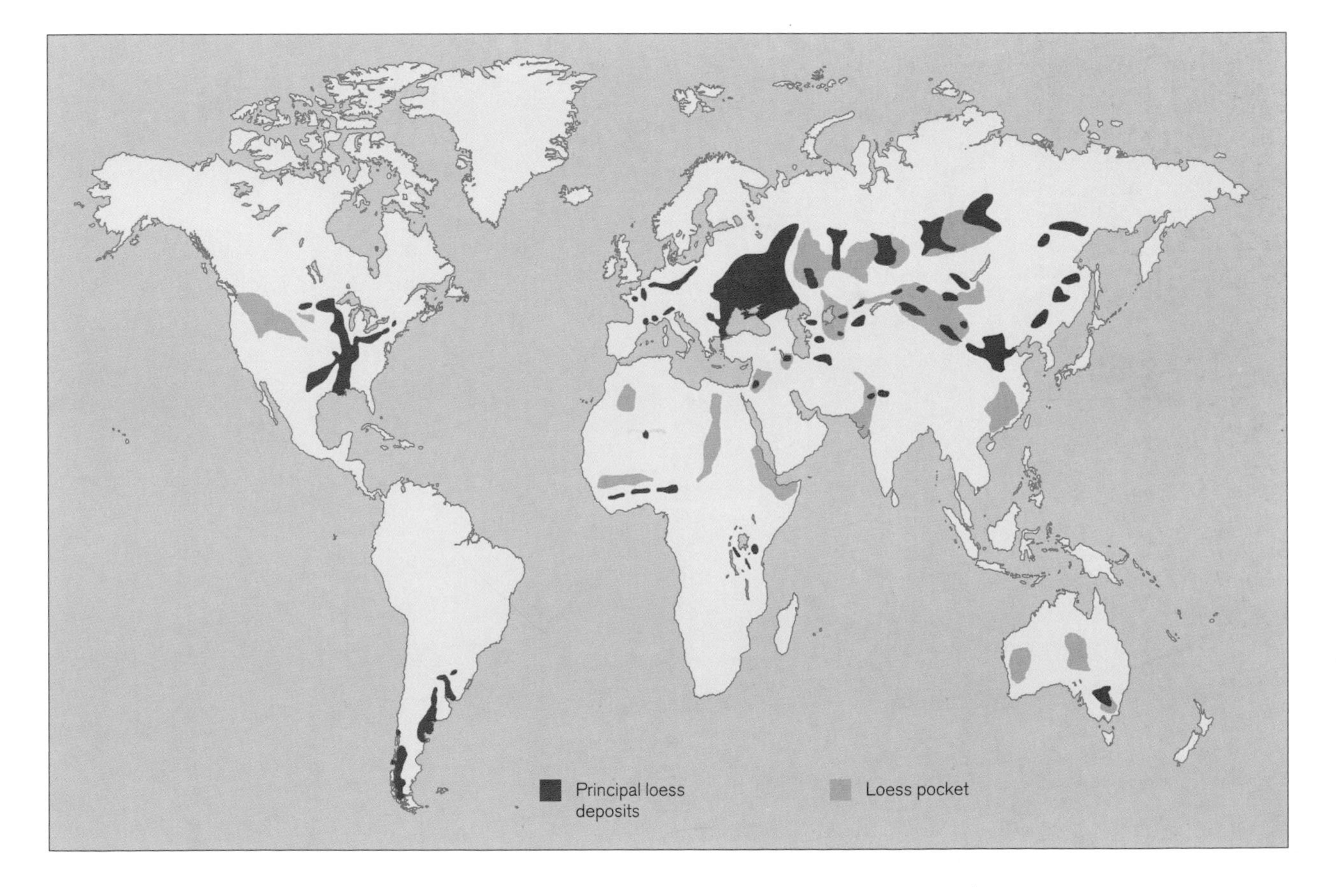

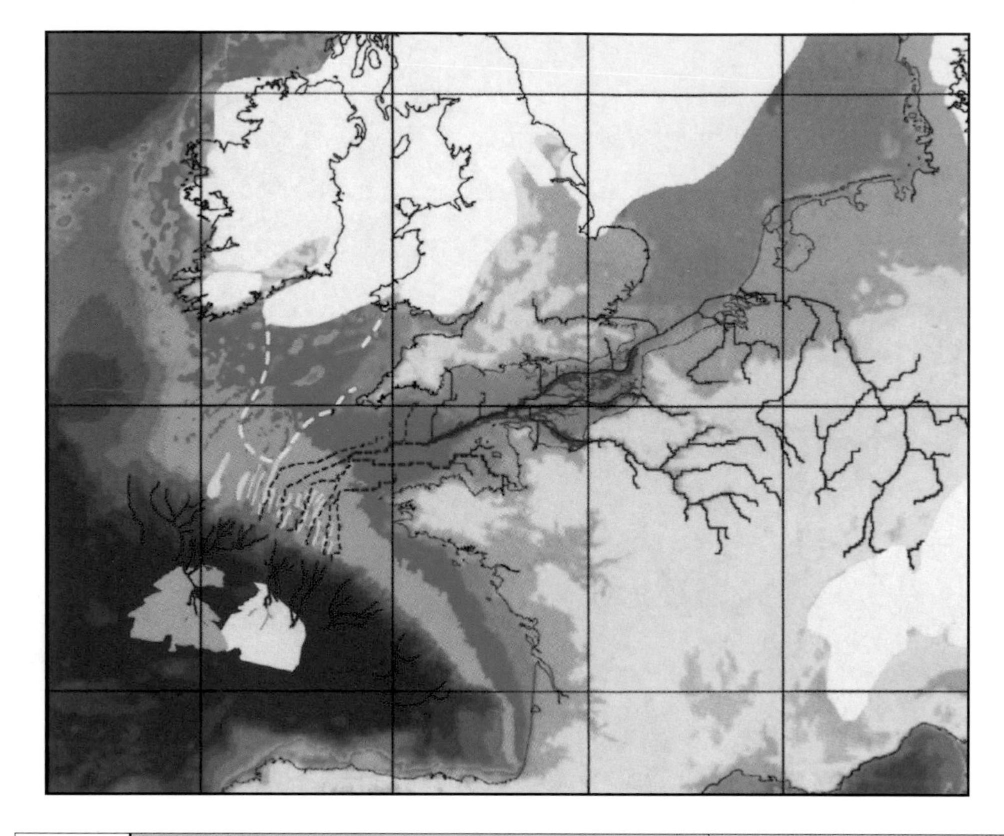

left During the last glaciation 18,000 years ago, global sea levels were 120 m (394 ft) lower than today. These levels left the North Sea and English Channel completely dry. There was instead a mega-river flowing out to the Atlantic Ocean combining the water discharge of the Thames, the Rhine and the Seine.

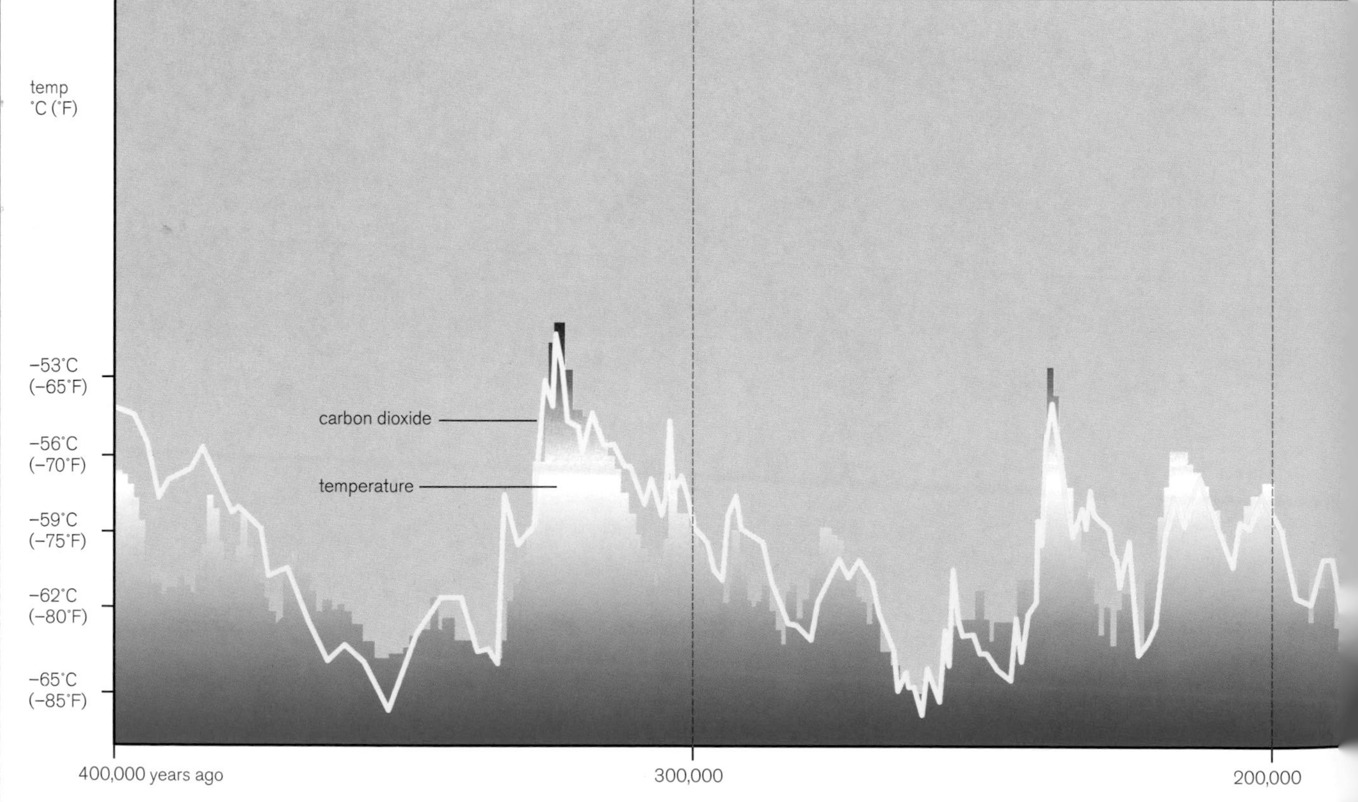

formed by the lowering sea levels, allowing species of all kinds of animal to invade new areas. Around the world, islands such as Sri Lanka, Japan, Sicily, Papua New Guinea and the Falklands became part of the adjacent mainland. Significantly, the chain of islands across the Bering Sea, which separates northeast Asia from Alaska, also became joined, so that during the end of the last glacial period as the climate started to warm up humans were able to cross from Asia into North America for the very first time, colonizing a brave new world (see Chapter 5).

The Ice Age also affected the atmosphere: carbon dioxide, as we have seen, was reduced by a third and methane by a half. This was due to massive changes in the carbon system. Methane, for example, was greatly reduced because of the drier conditions during the glacial periods. At the moment a lot of natural methane is produced in tropical wetlands when plants rot under water – in the Amazon, an area the size of Britain is flooded every year – so in a drier world less methane is produced. Both carbon dioxide and methane are important greenhouse gases so their reduction allowed more of the heat generated through solar radiation to escape the atmosphere, helping to cool the planet. As we will see later in this chapter, this factor may actually have been instrumental in the waxing and waning of the ice sheets. Finally, the total weight of plants on the land could have been reduced by as much as half.

Carbon dioxide is an important greenhouse gas, which we know from ice cores was reduced by a third during each glaciation, helping to make these periods even colder. On the graph below, the record of carbon dioxide is shown as a yellow line, while temperatures over Antarctica are shown in blue, with the warmest peaks indicated in red. The graph clearly demonstrates the powerful correlation between carbon dioxide and temperature. The alarming increase in carbon dioxide at the present may presage significant global warming.

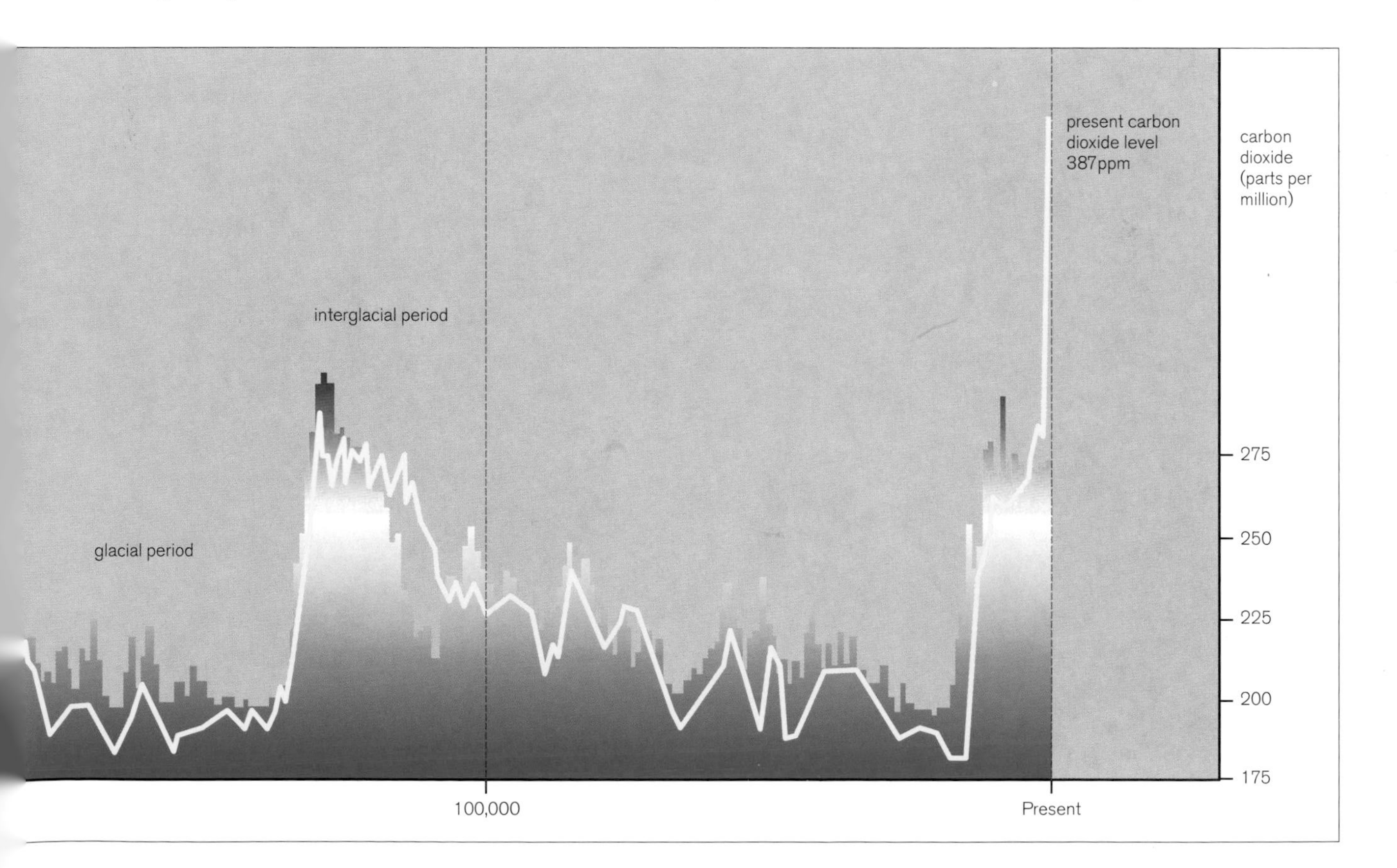

Case of the missing grass in Amazonia

Ice ages clearly affect the whole of the global climate system; however, there is controversy over what effects they had on the tropics. Half the surface of the planet lies between the Tropics of Cancer and Capricorn and this area includes all the tropical rainforests of the world. Of these the most important region in terms of size and species diversity is the Amazon. The Amazon Basin is the largest river basin in the world, covering an area of 7 million sq. km (2.7 million sq. miles), and it discharges approximately 20 per cent of all freshwater carried to the oceans. The majority of the Amazon Basin is covered by extremely diverse rainforest. In 1969 the German geologist Jürgen Haffer put forward a wonderful theory linking the Ice Age to the question of why the Amazon was so diverse. During each glacial period, he suggested, lower temperatures and precipitation in the tropics allowed savannah to replace the majority of the tropical rainforest. However, some of the tropical rainforest would have survived in small refugia, isolated islands of rainforest surrounded by grassland. These isolated patches of rainforest would have become hotbeds of evolution, producing many new species. At the end of each glacial period the patches of rainforest, each with its with higher level of species diversity, would

The Amazon rainforest (*opposite*) is one of the most species-rich regions in the world and scientists have put forward many theories about why it is so diverse. It was originally thought that during each glaciation most of the Amazon was reduced to savannah (*left*), which would cause evolutionary stress, generating increased diversity. We now know the rainforest is much more resilient and its area may have been reduced by just 20 per cent during the last glaciation, so new theories have to be put forward for the diversity of this remarkable ecosystem.

merge back together. By the late 1990s this theory came under attack, as more and more scientists failed to find evidence of a huge increase in savannah.

We now know from pollen records and computer models that in the Amazon the combination of the dry and cold conditions meant that savannah did encroach a little bit at the edges, reducing the area of the Amazon rainforest to 80 per cent of today's coverage, but it is a testament to the resilience and importance of tropical rainforest in the global ecosystem that the Amazon survived and even flourished during glacial periods. One of the reasons for this was that the cold conditions actually helped to reduce the problem of less rainfall by reducing the amount evaporation from the trees and thus the loss of essential moisture. However, there were major changes in the species composition of the Amazon rainforest during glacial periods. We know, for example, from pollen records that many of the tree species now found in the Andes were once found in the heart of the Amazon forest. This is because the species that are more cold-adapted are pushed up to higher, colder altitudes during warm interglacials. We cannot, therefore, see the current Amazon rainforest as the area's 'normal' condition, because for the last million years the earth's climate has spent about 80 per cent of the time in glacial conditions. The geological evidence shows us that the Amazon forest during last glacial period had a diverse mix between Andean and lowland tropic tree species. The lack of grassland in the Amazon during glacial periods also means we have to look for other evolutionary mechanisms for the huge diversity of the Amazon rainforest, and it may be that the glaciations were not the cause.

Waxing and waning of the great Ice Age

We now know that glacial–interglacial cycles are the fundamental characteristic of the last 2.5 million years. The waxing and waning of the huge continental ice sheets, described above, is initiated by the changes in the earth's orbit around the sun. Over long periods of time the earth wobbles on its axis and thus changes the amount of sunlight or solar energy received by different parts of the earth. These small changes are enough to push or force climate to change, but these waxings and wanings are not directly caused by the earth's orbital wobbles, rather they are caused by the earth's climatic reaction, which translates relatively small changes in regional solar energy into major climatic variability. To illustrate this idea, we need only consider the fact that the position of the earth today is very similar to what it was 21,000 years ago during the last glacial period, when ice lay over 2 km (1.2 miles) deep over North America. It is not, therefore, the exact orbital position that controls climate but rather the *changes* in the orbital positions. There are three main orbital parameters or wobbles, called eccentricity, obliquity (tilt) and precession (see boxes), and each has

Eccentricity

Eccentricity is the changing shape of the earth's orbit from near-circular to an ellipse over a period of about 96,000 years. Imagine a new rubber band. When you drop it on the table it makes a perfect circle. If you put your two fingers into it and gently pull apart you make an ellipse. The orbit of the earth goes from that circle to the ellipse and back again every 96,000 years. Described another way, the long axis of the ellipse varies in length over time. Today the earth is at its closest (146 million km or 91 million miles) to the sun – a position is known as the perihelion – on 3 January. On 4 July it is at its most distant from the sun (156 million km or 97 million miles) – this position is called the aphelion. Changes in eccentricity cause only very minor variations in the total annual solar radiation or energy (approximately 0.03 per cent), but can have significant seasonal effects. If the orbit of the earth were perfectly circular there would be no seasonal variation in solar energy. Milankovitch suggested in 1949 that the Northern Hemisphere ice sheets are more likely to form when the sun is more distant in summer, so that each year some of the previous winter's snow can survive. The intensity of solar radiation reaching the earth diminishes in proportion to the square of the planet's distance, so modern global insolation is reduced by nearly 7 per cent between January and July. As Milankovitch suggested, this produces a situation that is favourable for snow accumulation; however, it is more favourable for snow surviving in the Northern rather than the Southern Hemisphere. The more elliptical the shape of the orbit becomes, the more the season will be exaggerated in one hemisphere and moderated in the other. The other effect of eccentricity is to modulate the effects of precession, which, as we will see, has a much stronger influence than eccentricity. It is essential to note that eccentricity is by far the weakest of the three orbital parameters.

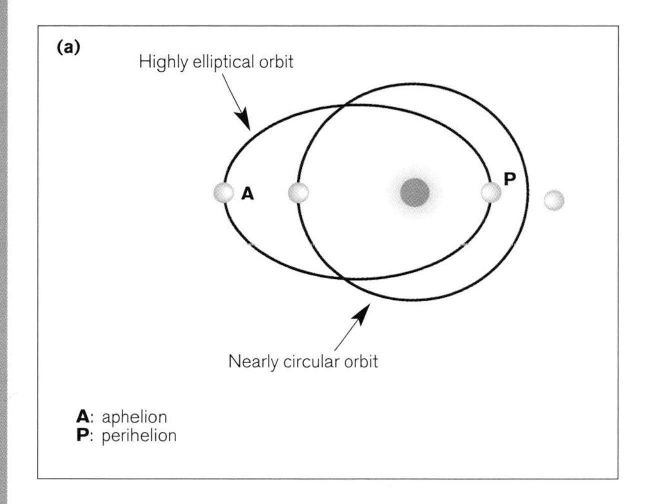

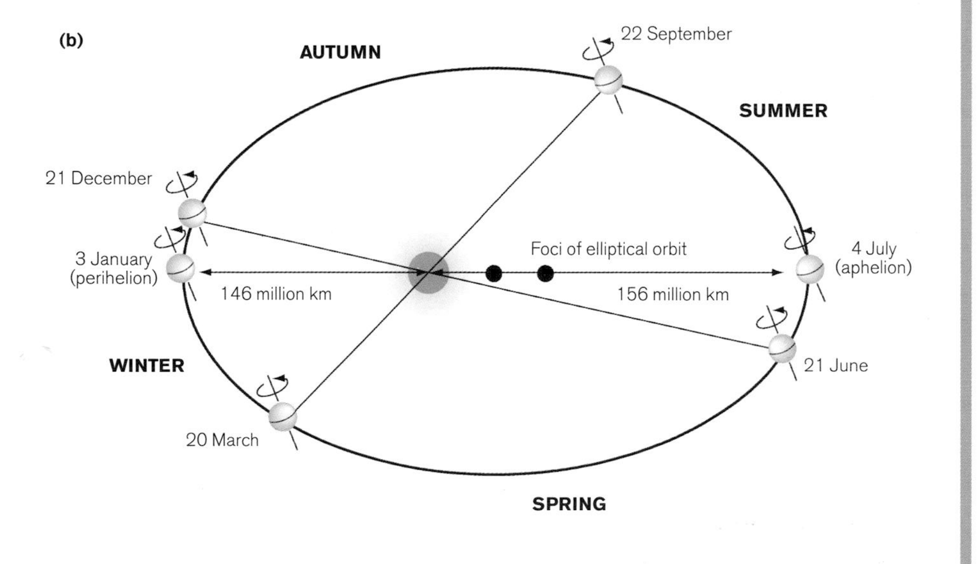

Changes in the shape of the earth's orbit around the sun. Diagram (a) shows how the shape of the orbit changes from near-circular to elliptical. The position along the orbit when the earth is closest to the sun is termed the perihelion and the position when it is furthest from the sun is termed aphelion. Diagram (b) shows the present-day orbit and its relationship to the seasons, and shows the earth is closer to the sun in Northern Hemisphere winter than summer.

Obliquity

The tilt of the earth's axis of rotation with respect to the plane of its orbit (the plane of the ecliptic) varies between 21.8° and 24.4° over a period of 41,000 years. It is the tilt of the axis of rotation that gives us the seasons, since in summer the hemisphere that is tilted towards the sun is warmer because it receives more than 12 hours of sunlight, and the sun is higher in the sky. At the same time, in the opposite hemisphere, the axis of rotation is tilted away from the sun and so that region is plunged into winter, being colder because it receives less than 12 hours of sunlight, and the sun is lower in the sky. Hence, the larger the obliquity, the greater the difference between summer and winter. As Milankovitch suggested, the colder the Northern Hemisphere summers the more likely ice sheets are to build up. This is why there seems to be a straightforward explanation for the glacial–interglacial cycle occurring every 41,000 years prior to 1 million years ago. Despite the changes in solar energy at any circle of latitude being dominated by precession, Professor Maureen Raymo (Boston University) has shown that obliquity controls the heating exchange between high and low latitudes. Hence the atmospheric north–south flux of heat and moisture, which exerts a dominant control on global climate, varies every 41,000 years. This is because the majority of heat transport between 30°N and 70°N occurs by way of the atmosphere: thus a linear relationship between obliquity, northward heat transport and glacial–interglacial cycles can be envisioned.

a unique cycle and effect on climate. More exciting, though, is when we combine them and see how together they push the climate either into or out of an Ice Age.

Clockwork climate

The boxes describe each of the orbital parameters. By combining the effects of all three – eccentricity, obliquity and precession – we can calculate the solar energy received for any latitude back through time. Milutin Milankovitch suggested in 1949 that the solar radiation received in summer (summer insolation) at 65°N was critical in controlling glacial–interglacial cycles. He argued that if summer insolation was reduced enough then ice could survive through the summer and thus start to build up, eventually producing an ice sheet. Orbital forcing does have a large influence on this summer insolation – the maximum change in solar radiation in the last 600,000 years is equivalent to reducing the amount of summer radiation received today at 65°N to that received now over 550 km (340 miles) to the north at 77°N. In simplistic terms this brings the current ice limit in mid-Norway down to the latitude of mid-Scotland. These lows in 65°N insolation are caused by eccentricity elongating the summer earth–sun distance, obliquity being shallow and precession placing the summer season at the longest earth–sun distance produced by eccentricity. The reason why it is 65°N and not 65°S that controls climate is very simple: any ice that builds up in the Northern Hemisphere has lots of continents to grow upon. In contrast, ice growth in the Southern Hemisphere is limited by the Southern Ocean around Antarctica, since any extra ice produced on Antarctica falls into the ocean

Precession

There are two components of precession: that relating to the elliptical orbit of the earth and that related to its axis of rotation. The earth's rotational axis moves around a full circle, or 'precesses', every 27,000 years. This is similar to the gyrations of the rotational axis of a toy spinning top. When the spinning top is spinning around its plunger the plunger also rotates, but at a much slower speed. In the same way the earth spins round once every day but the axis of rotation, the earth's 'plunger', takes much longer (about 27,000 years) to rotate once. Precession causes the dates of the equinoxes to travel around the sun, resulting in a change in the earth–sun distance for any particular date. The combined effect is illustrated in Figure c, which shows the precession of the earth's orbit, which has a periodicity of 105,000 years, and how it changes the time of year when the earth is closest to the sun (perihelion). Imagine a child spinning a hula hoop slowly around one foot. If the hula hoop represents the earth's orbit around the sun, then the spinning round the foot represents the precession of the earth's orbit. It is the combination of the different orbital parameters that results in the classically quoted precessional periodicities of 23,000 and 19,000 years. Combining the precession of the axis of rotation and the precessional changes in orbit produces a period of 23,000 years. Combining the shape of the orbit i.e. eccentricity and the precession of the axis of rotation results in a period of 19,000 years. These two periodicities combine so that perihelion coincides with the summer season in each hemisphere on average every 21,700 years, resulting in the precession of the equinoxes. Precession has the most significant impact in the tropics (in contrast to the impact of obliquity at the Equator, which is zero). So although obliquity clearly influences high latitude climate change, which may ultimately influence the tropics, direct effects of insolation in the tropics are due to eccentricity-modulated precession alone.

The components of the precession of the equinoxes. Diagram (a) shows the precession of the earth's axis of rotation, which takes 27,000 years to wobble round once. Diagram (b) shows the precession of the earth's orbit around the sun, which takes 105,000 years to wobble round once. Diagram (c) shows the precession of the equinoxes, and when the earth is closest to the sun.

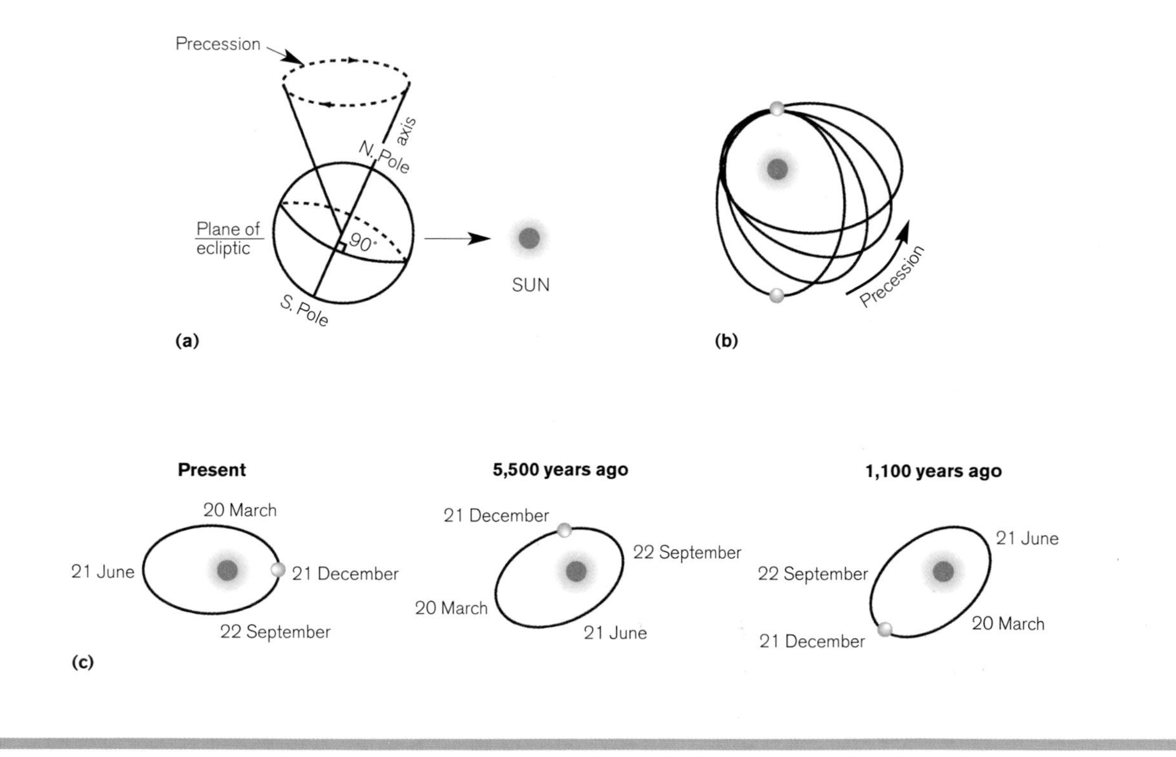

and is swept away to warmer seas. So the conventional view of glaciation is that low summer solar energy in the temperate Northern Hemisphere allows ice to survive through summer and so ice sheets start to build up on the northern continents. In reality this clockwork view of the world is far too simple, since the effects of orbital changes on the seasons are very small and, as we will see, it is actually feedback mechanisms in the climate system that amplify these changes.

What causes ice ages?

Orbital forcing in itself is insufficient to drive the observed glacial–interglacial variability in climate. Instead, the earth system amplifies and transforms the changes in solar energy received at the earth's surface through various feedback mechanisms. Let us start, for example, by looking at the systems that generate a glacial period. The first thing to happen when solar radiation is reduced is a slight reduction in summer temperatures. As snow and ice accumulate due to these initial changes in summer temperature, the ambient environment is modified – primarily by an increase in albedo, the reflection of sunlight back into space. Just think of having to wear sunglasses when skiing because the snow reflects most of the sunlight that falls on it straight into your face. Reflecting more sunlight back into space suppresses local temperatures, which promotes the accumulation of more snow and ice, thus generating a further modification of the ambient environment – the so-called 'ice-albedo' feedback. So once you have a small ice sheet it changes the environment around it to make more snow and ice and will get bigger and bigger.

Another feedback is triggered when the ice sheets, particularly the Laurentide ice sheet on North America, become high enough to deflect the jet stream which controls the position of the warm and cold fronts of air that move across the North Atlantic Ocean towards Europe. This changes the storm path across the North Atlantic and prevents Gulf Stream and North Atlantic Drift from penetrating as far north as they do today. This surface ocean change, combined with the general increase in meltwater in the Nordic Seas and Atlantic Ocean due to the presence of large continental ice sheets, ultimately leads to a reduction in the production of deep-water. As we saw in Chapter 3, deep-water production in the Greenland and Labrador Seas is the heartbeat of modern climate. Reducing the formation of deep-water reduces the amount of warm water pulled northwards, all of which leads to increased cooling in the Northern Hemisphere and expansion of the ice sheets.

However, there is currently a debate among palaeoclimatologists about the primacy of the role of the 'physical climate' feedbacks just described over the role of greenhouse gases in the atmosphere. A reduction in the atmospheric concentration of greenhouse gases, such as carbon dioxide, methane and water vapour, will drive a general global cooling. The reduction in carbon dioxide and methane that occurred during each glacial period has already been shown in air bubbles trapped in polar ice, and these cold episodes are also by their very nature drier, which reduces atmospheric water vapour. So the debate continues: do changes in the earth's orbit affect the production of greenhouse gases, thus cooling down the earth and making the Northern Hemisphere susceptible to the build-up of large ice sheets, or do changes in the earth's orbit start to build up large ice sheets in the Northern Hemisphere, which in turn changes global climate and reduces the production of greenhouse gases, thus prolonging and deepening the glacial period? The jury is still out on this one. Either way, however, greenhouse gases certainly played a critical role in producing regular glacial–interglacial cycles.

Another important question to consider is: why don't these feedbacks end up running away and freezing the whole earth? They are prevented from doing this by 'moisture limitation'. To build an ice sheet the climate needs to be cold and wet. However, as we have seen, when ice sheets form, they alter the circulation of warm and cold air and water around the globe. As warm surface water is forced further and further south, the supply of moisture that is required to build ice sheets decreases. So the ice sheets, by changing the atmospheric and oceanic circulation, end up starving themselves of moisture and thus limiting their own growth.

In the last million years it took up to 80,000 years for the ice sheets to build up to their maximum extent. The last time this occurred was about 21,000 years ago. Getting rid of the ice is a much quicker process. Called deglaciation, it usually took

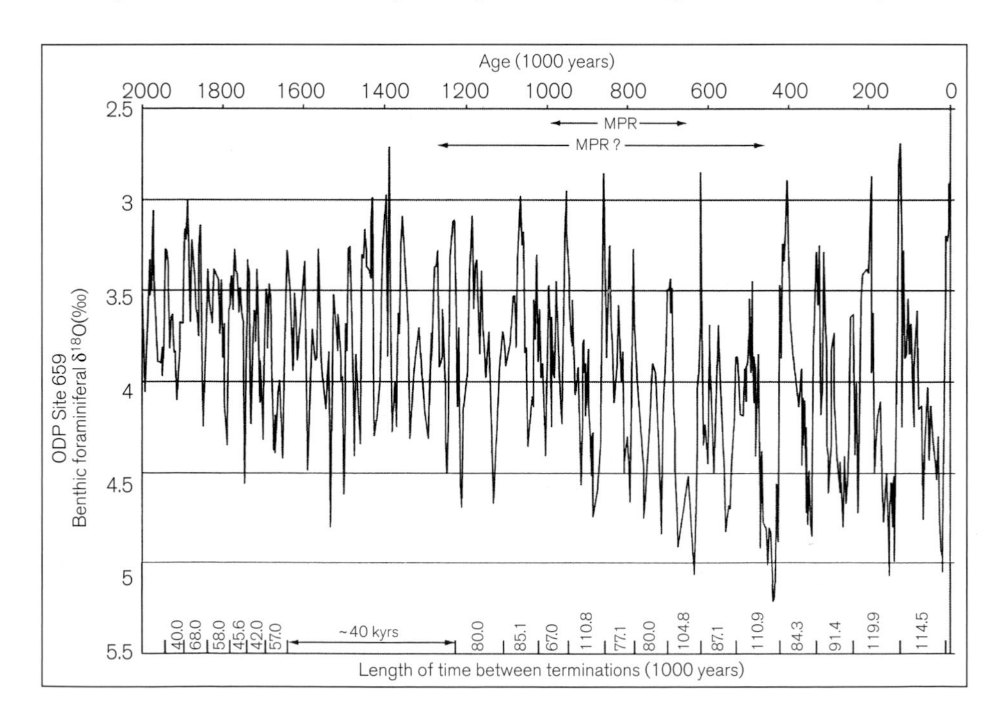

This oxygen isotope record from the deep ocean, showing the changes in the amount of ice on earth, reveals the timing of the glaciations of the Ice Age. Before 1 million years ago these glaciations occurred approximately every 41,000 years, after which point they started to become longer. This climate transition is called the Mid-Pleistocene Revolution (MPR) but scientists are unsure how long this transition should continue for.

a maximum of only 4,000 years. Deglaciation is triggered by an increase in solar energy received in the summer at about 65°N, which encourages the Northern Hemisphere ice sheets to melt slightly. As the world gets wetter and ocean circulation changes, both atmospheric carbon dioxide and methane start to rise, promoting warming globally and encouraging the melting of the large continental ice sheets. These processes are limited, however, because they have to work against the ice sheets' albedo effect, which produces a microclimate that tries to keep them intact.

The climate during a glacial period was very dynamic and constantly changing. This is because ice is naturally unstable and ice sheets continually calve into the oceans and can even completely collapse.

What causes the rapid removal of ice is actually the rise in sea level due to the initial melting of the ice, as large ice sheets adjacent to the oceans are undercut by the sea. The coldest sea water can be is about –1.8°C (29°F), while the base of the ice sheet is usually colder than –30°C (–22°F), so it is like putting hot water under a tub of ice cream. This undercutting of the ice sheet therefore leads to more melting and ice subsequently calving into the ocean, which increases the sea level, causing even more undercutting. This sea-level feedback mechanism can be extremely rapid. Once the ice sheets are in full retreat then the other, ice-forming feedback mechanisms discussed above are thrown into reverse.

The Ice Age climate rollercoaster

We have called this chapter 'The Climatic Rollercoaster' because ice sheets are naturally unstable and during glacial periods the climate veers violently from one state to another as the ice sheets dramatically collapse and then reform. Most of the variations occur on the millennial timescale, but as we will see they can start in as little as three years.

The most impressive of these climate change events are the so-called Heinrich events, named by Wallace Broecker of Columbia University after he saw a paper by the marine geologist Hartmut Heinrich describing them in 1988. Heinrich events are massive collapses of the North American Laurentide ice sheet, which resulted in millions of tons of ice being poured into the North Atlantic Ocean. Broecker described them as 'armadas of icebergs' floating from North America across the

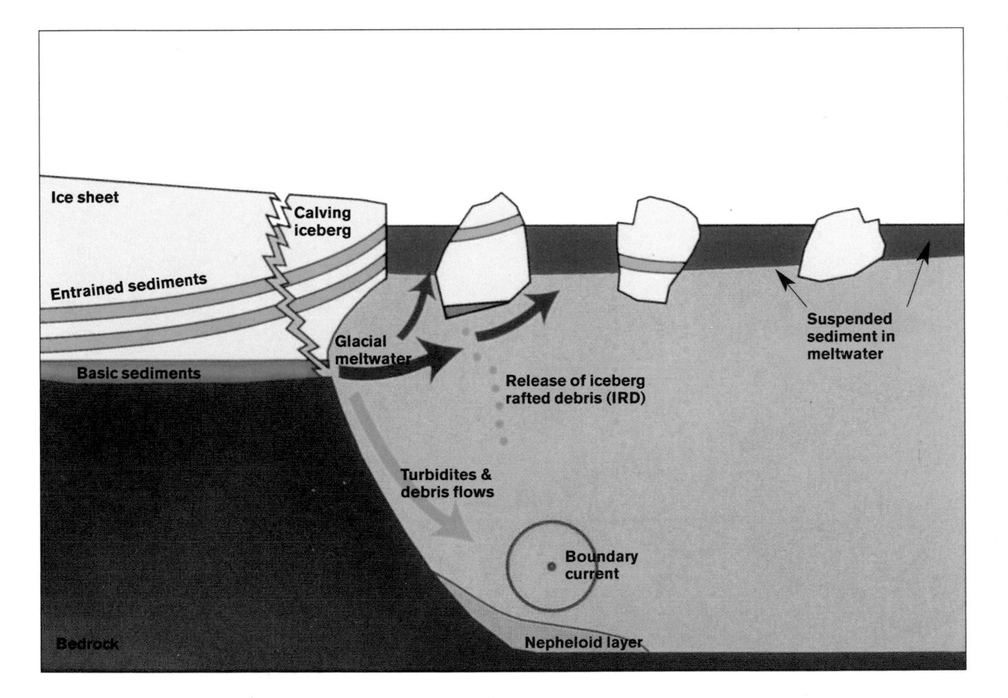

Icebergs carry a huge amount of sediment within them when they calve from ice sheets. As the icebergs melt they drop this sediment as a trail of debris across the ocean floor (known as the nepheloid layer) which scientists can follow to work out where the icebergs flowed to.

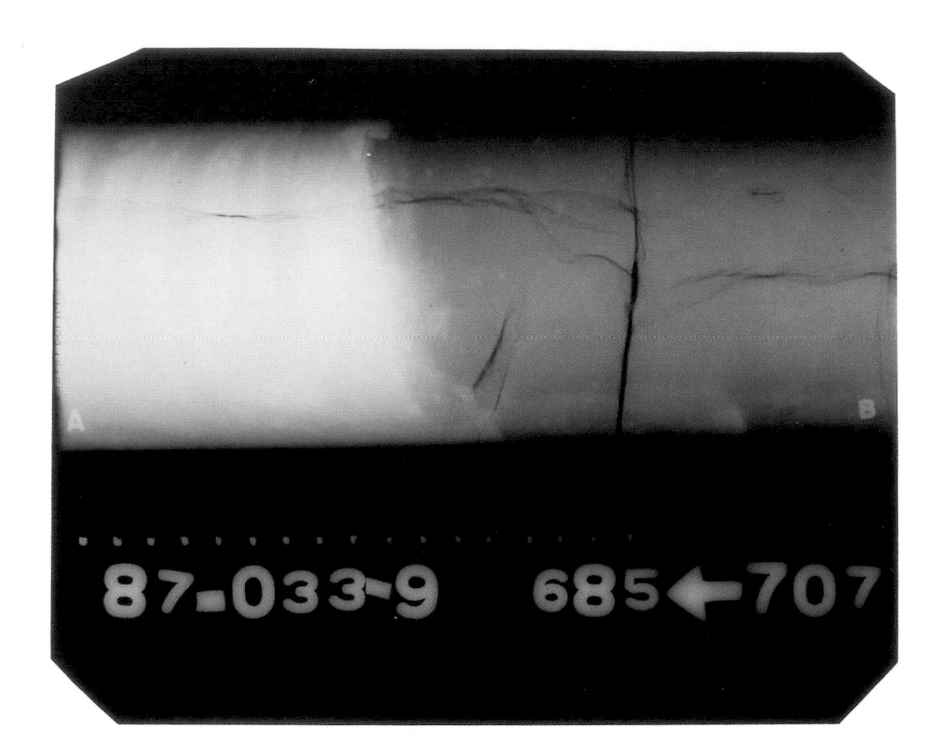

left An X-ray picture of a marine sediment core showing the boundary between the fossil-rich normal Ice Age sediment (which shows up as dark material on the right of the X-ray) and the iceberg-debris-rich sediment in a Heinrich layer (light material on the left).

below An ocean core taken from the North Atlantic Ocean which has penetrated the top of a Heinrich event. The fossil-rich sediment is the lighter material on top and the iceberg-debris-rich sediment of the Heinrich layer is below.

Atlantic Ocean to Europe, and huge gouges have been found on the north French coast where these great icebergs ran aground. Heinrich events occurred against the general background of an unstable glacial climate, and represent the brief expression of the most extreme glacial conditions around the North Atlantic. They are evident in the Greenland ice core records as a further 3–6°C (5–11°F) drop in temperature from the already cold glacial climate. Heinrich events have also been found to have had a global impact, with evidence for major climate changes described from as far afield as South America, the North Pacific, the Santa Barbara Basin, the Arabian Sea, China, the South China Sea and the Sea of Japan. While these events were occurring around the North Atlantic region much colder conditions are found both in North America and Europe. This is because in the North Atlantic Ocean the huge number of melting icebergs added so much cold fresh water that the sea surface temperatures and salinity were reduced to the point where surface water could not sink, thereby stopping all deep-water formation in the North Atlantic Ocean, and switching off the global ocean conveyor belt that circulates warm water from the tropics.

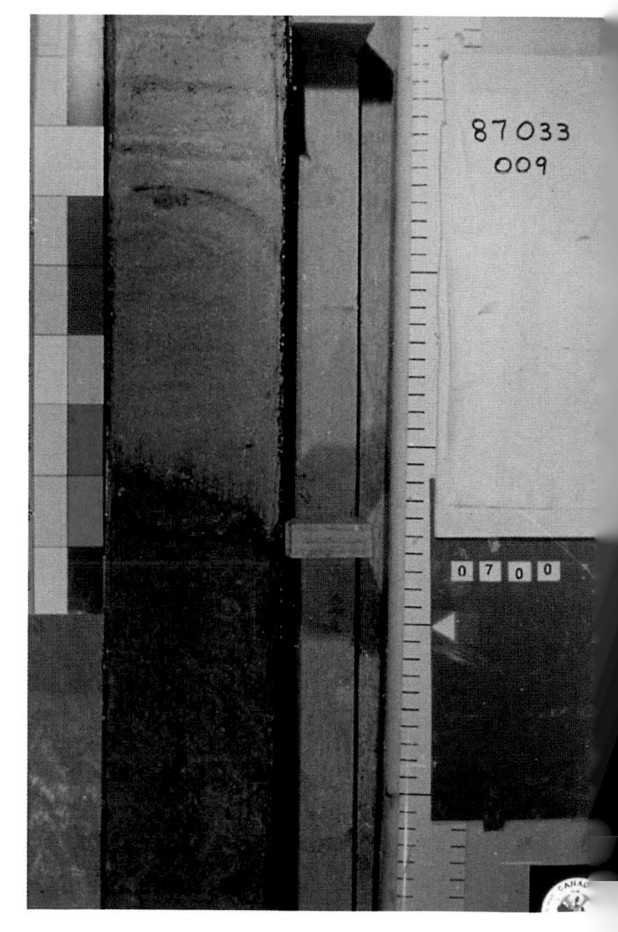

What Caused the Heinrich Events?

Heinrich events are fascinating because they occur on a timescale that we can appreciate and had a massive and profound effect on the climate. There are, therefore, lots of competing theories about what caused them. The glaciologist Doug MacAyeal suggested that the Heinrich event iceberg surges were caused by internal instabilities in the Laurentide ice sheet. This ice sheet rested on a bed of soft, loose sediment, which behaves like concrete when it is frozen, and so would have been able to support the weight of the growing ice sheet. As the ice sheet expanded, the geothermal heat from within the earth's crust, together with heat released from friction of ice moving over ice, was trapped by the insulating effect of the overlying ice. This 'duvet' effect allowed the temperature of the sediment to increase until a critical point when it thawed. When this occurred the sediment became soft, and thus lubricated the base of the ice sheet, causing a massive outflow of ice through the Hudson Strait into the North Atlantic. This, in turn, would have led to sudden loss of ice mass, which would have reduced the insulating effect and caused re-freezing of the basal ice and sediment bed, at which point the ice would have reverted to a slower build-up and outward movement. Doug MacAyeal called this a 'binge–purge' model and suggested that all ice sheets have their own instability times, thus the Scandinavian, Greenland and Icelandic ice sheets would have surges with different periodicities.

Another exciting theory is the 'bipolar climate seesaw' idea – another wonderful term created by Wally Broecker. This theory is based on new evidence from ice cores in Greenland and Antarctica, which shows that during Heinrich events the climates of the Northern and Southern Hemispheres are out of phase, so that when the climate is cooling down in the Northern Hemisphere it is warming up in the Antarctic. It has been suggested that this bipolar climate seesaw can be explained by alternating ice sheet collapse and resultant meltwater surges in the North Atlantic and Southern oceans. Each meltwater surge would change the relative amount of deep-water formation in the two hemispheres and the resulting direction of the inter-hemispheric heat piracy.
At the moment the Northern Hemisphere steals heat from the Southern Hemisphere to maintain the Gulf Stream and the formation of relatively warm deep-water in the Nordic Seas. The heat is slowly returned by the flow of deep-water from the North to the South Atlantic Ocean. So the bipolar climate seesaw model suggests that if the ice sheets around the North Atlantic collapsed, sending huge numbers of icebergs into the ocean, these icebergs would melt, making the ocean so fresh that none of the water could sink. This would stop the formation of North Atlantic deep-water and the Northern Hemisphere would stop stealing heat from the Southern Hemisphere, resulting in the Southern Hemisphere slowly warming up. Over maybe 1,000 years this heat build-up would be enough to collapse the edges of the Antarctic ice sheet, which would then shut off the deep-water formation around Antarctica and the whole system would be reversed. The nice thing about this theory is that it can work in an interglacial as well, and, as we see below, the Dansgaard–Oeschger cycles of about 1,500 years occur during both glacial and interglacial periods (see also Chapter 7).

Heinrich events are easy to spot in marine sediment cores from the middle of the Atlantic Ocean. This is because the icebergs bring huge amounts of rock with them into the ocean and as they melt they leave a trail of rock fragments scattered over the floor of the ocean. By recognizing these events in marine sediments and dating the fossils in the sediment, it seems that the Heinrich events occurred on average every 7,000 years during the last glacial period. In addition to the fossil evidence, below

The compositional difference between normal Ice Age marine sediment (bottom right) and sediment deposited during a Heinrich event (top left) can be clearly seen here. The marked difference occurs because during Heinrich events vast quantities of rock material are brought into the ocean from the continent by icebergs and are deposited on the sea bed.

these rock fragment horizons we have found little burrows of marine worms – usually these burrows cannot be seen as the sediment is mixed up by other animals coming to feed on it. For these fossil tubes and burrows to be preserved, the rain of rock fragments from the melting icebergs must have occurred within three years and been rapid enough to prevent other animals getting to the sediment. This evidence suggests that the collapse of the North American ice sheet was extremely rapid, with icebergs flooding the Atlantic Ocean likewise in less than three years. So during glacial periods conditions varied from cold conditions with massive ice sheets to extreme cold conditions brought on by the partial collapse of the North American ice sheet.

We now know that between the massive Heinrich events there are smaller events occurring at about every 1,500 years, which are referred to as Dansgaard–Oeschger cycles. These are also caused by meltwater being pumped into the North Atlantic Ocean. One suggestion is that Heinrich events are in fact just super Dansgaard–Oeschger cycles. The big difference between the two types of event is that Heinrich events are found only during glacial periods, while Dansgaard–Oeschger cycles have been found in interglacials as well as glacials. In fact, as we will see in Chapter 7, there were six major Dansgaard–Oeschger cycles (known in the Holocene as 'Bond events') in the last 10,000 years, including one 4,200 years ago that caused a series of massive droughts in the Middle East and may well have radically affected ancient civilizations at the time.

Conclusion

Over the past 2.5 million years the earth's climate has been in a constant state of flux as it cycles between glacial and interglacial periods. Between 2.5 million and 1 million years ago the glacial periods came and went every 41,000 years, driven by changes in orbital obliquity that were dramatically amplified by climatic feedback systems. After 1 million years ago the glacial periods intensified and lasted about 100,000 years; if we look at a graph showing climatic changes we can see that at this point it becomes saw-toothed, with a long period of ice build-up usually lasting at least 80,000 years, followed by a dramatic rebound back to a warm interglacial climate in less than 4,000 years.

The switches between glacial and interglacial periods have not been the only climate change events to occur during the last Ice Age, however. During the intense glacial periods the climatic rollercoaster continued with the huge ice sheet on North America regularly collapsing into the ocean. The vast armadas of icebergs that resulted from these collapses filled the North Atlantic and had a profound effect on global climate, plunging the world into even colder conditions. Most impressive is how rapidly these collapses occurred, with evidence of icebergs filling the North Atlantic Ocean in less than three years. These so-called Heinrich events provide a stark warning of the possible extreme and rapid effects of climate change in the future.

We do not always realize that in the temperate latitudes almost all the landscape around us has been touched by the Ice Age, from the shapes of mountains and valleys to the current position of the river Thames. The climatic rollercoaster has shaped our planet in myriad ways, and will continue to do so.

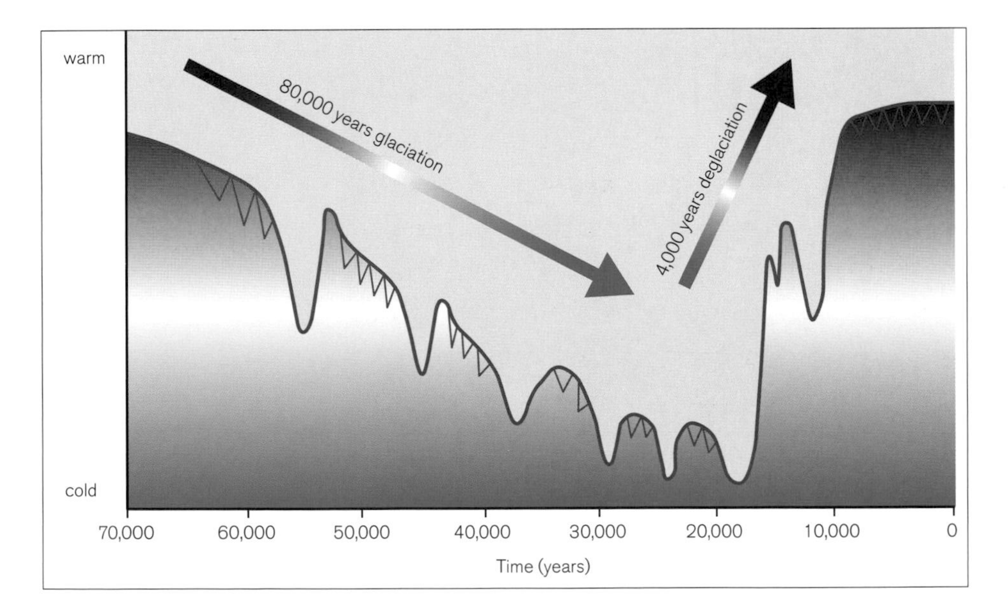

It takes a long time – about 80,000 years – for the climate to be pushed into a glacial period, but only about 4,000 years to bounce out of one. This is because of the unstable nature of ice sheets, which continually collapse even though the climate continues to get colder and colder. Each one of the large dips in temperature on the graph during the last glacial period represents a Heinrich event, while the small saw-toothed events that occur both during the glacials and interglacials represent the Dansgaard–Oeschger climate cycles.

5

The Human Story

HUMANS EVOLVED OUT OF THE AFRICAN APES as a distinct group of living beings more than 5 million years ago. At the beginning of the Ice Age – slightly less than 2 million years ago – they expanded out of Africa for the first time. With brains a little larger than those of the living apes and a set of simple tools, they invaded a variety of habitats in the middle latitudes and were living in places somewhat cooler, more seasonally variable and less rich in plant and animal foods than their tropical homeland. By the end of the Ice Age (roughly 12,000 years ago), humans had become the dominant animal on the planet. They were living in virtually every terrestrial habitat on earth and beginning to invade the oceans as well.

The transformation of humankind during the Ice Age lies above all in the emergence of the human mind. All higher forms of animal life possess a brain, which processes information about the surrounding environment. Only humans have developed the unique phenomenon of the mind, which does not simply process information, but generates it as well. With the creative power of the mind, and the implementing power of their hands, humans reshaped themselves and their surroundings in countless ways, adapting to – and thriving in – places where their ancestors would have perished in hours.

The mind is the centrepiece of the human story and most of the events critical to its emergence took place during the Ice Age, which provides the setting for this story. It is now apparent that novel forms of humankind evolved repeatedly in Africa and migrated into Eurasia during the course of the past 2 million years. As each new form of human moved out of Africa, it confronted the effects of Ice Age climate change on the environments of the Northern Hemisphere. Those changes became increasingly extreme, as intervals of warmth – sometimes warmer than the present day – alternated with episodes of intense cold. The latter witnessed not only the growth of immense glaciers, but also created landscapes that have no parallel today – places where animals of the arctic tundra lived alongside those of the steppe and woodlands. Moreover, climate change during the Ice Age often occurred with breathtaking speed, forcing people to change their way of life without protracted debate.

Before the Ice Age

The human fossil record is extremely sparse prior to about 3.5 million years ago, and much of what we think we know about human origins is based on the measure of genetic difference between ourselves and our closest living relatives. Because genetic change over time occurs at a regular pace, the amount of genetic difference between two species is a rough measure of the time elapsed since they were one. The genetic differences between ourselves and the African apes are relatively small and indicate an evolutionary split only a few million years ago.

previous pages Humans evolved in the tropics, but spread north into higher latitudes during the Ice Age and were forced to adapt to conditions of extreme cold. The Neanderthals (inset) evolved some physical adaptations to cold, including thick chests and shortened extremities, but also adjusted to Ice Age environments with a diet high in animal protein and fat and heavy use of fire.

When this was first revealed in the 1960s, it shocked anthropologists who believed at the time that the earliest humans were as much as 15 million years old. But the fossil record now supports this conclusion. Fossils that clearly represent an early form of human may be dated to 4 million years ago – these are the so-called australopithecines ('southern apes') – while more problematic specimens are dated to between 5 and 6 million years ago. Recent discoveries in central and east Africa include some even earlier fossils dating to between 6 and 7 million years ago, but their status as humans is unclear.

Olduvai Gorge in northern Tanzania, where Louis and Mary Leakey first discovered human remains of the genus *Homo* in the 1950s. This steep-sided ravine in the Great Rift Valley has yielded an astonishing number of finds crucial to our understanding of early human evolution.

The key anatomical feature used to identify these fossils as early humans is bipedalism – walking upright on the hindlimbs. Bipedalism is actually a complex of traits that may be seen in many parts of the skeleton. Anthropologists have searched intently among the isolated and fragmentary fossils for clues that point to bipedalism. It was the shift to upright walking that almost certainly produced humans out of apes. Moreover, bipedalism seems to underlie most or all of the important developments that followed. Walking upright freed the hands for making and using tools and weapons. This in turn was related to the increasing role of hunting and consumption of meat. Bipedalism may be indirectly connected to the growth of the brain and the evolution of the human vocal tract.

Why did some African apes start walking on their hindlimbs? We may never know the answer for certain, but the current consensus among anthropologists is that bipedalism is tied to energy-efficient movement on open ground (as opposed to climbing trees in a forest). There is also widespread belief that walking evolved in response to shrinking forest and a corresponding expansion of savannah and open woodland. Like other parts of the world, Africa became cooler and drier after 8 million years ago, apparently due to growing glaciers at the poles. The African forests became smaller, creating new opportunities for animals that could cope with more open landscapes.

Footprints of australopithecines, preserved in volcanic ash at Laetoli in Tanzania, provide dramatic illustration of bipedal walking roughly 3.5 million years ago.

By 3.5 million years ago, the australopithecines were well established in east Africa. A comparatively rich fossil record – including footprints – provides a vivid picture of their appearance and habits. They have sometimes been described as 'bipedal apes', because they were ape-like in many respects, particularly in their small ape-sized brains. And although bipedal, they retained some ape-like limb characteristics as well, suggesting that they were still spending a good deal of time in the trees. Their diet was based on plant foods; there is no evidence of meat-scavenging or hunting. But while there are no recognized tools from this time range, the australopithecine hand exhibits some human-like features that foreshadow subsequent developments.

Roughly 2.5 million years ago, human evolution suddenly turned in a new direction. The first recognized stone tools appeared in east Africa. They were simple tools – pebble choppers and flakes for scraping and cutting – but experimental research shows that their manufacture is beyond the capacity of living chimpanzees. At least some of these tools were used for stripping meat off the bones of animals, evidence of which now appears for the first time. Much of human anatomy remained unchanged with some significant exceptions: the brain increased in size, rising above the ape level for the first time, and the morphology of the hand continued to develop.

The fossil record for this critical period of transition from the australopithecines to a more recognizable form of human is shrouded in confusion. Several new human species appeared at this time, including a robust form of australopithecine with larger chewing teeth and a slightly larger brain. Most anthropologists also recognize two different forms assigned to our genus *Homo*. One of these (*Homo habilis*) was first discovered by Louis and Mary Leakey in 1960 and possessed a brain that – while larger than that of the living apes – was very small by our standards. The other form of *Homo* (*Homo rudolfensis*) had a somewhat bigger brain, as well as larger jaws and teeth.

It has been difficult to figure out who was making the tools and consuming the meat, although most assume that one or both forms of *Homo* were doing at least some of both. What is significant is that a few hundred thousand years before the beginning of the Ice Age, humans evolved some novel traits – larger brains, more agile hands and an improved capacity to make tools – that provided them with the means to live in a much wider range of climate settings and environments. By 1.8 million years ago, humans were living well outside the tropical zone. And eventually they would develop the ability to cope with – and even thrive under – the most extreme Ice Age conditions.

Pebble choppers and other simple stone tools from Bed I at Olduvai Gorge represent the earliest known form of human technology. Experimental research reveals that their manufacture is beyond the capacity of living apes.

Dmanisi is located on the southern side of the Caucasus Mountains in the Republic of Georgia. It contains the oldest known traces of humans outside Africa, dating to about 1.8 million years ago.

Out of Africa

Roughly 80 km (50 miles) southwest of the city of Tbilisi in the Republic of Georgia, beneath the ruins of a medieval castle overlooking the confluence of two rivers, lies one of the most important archaeological sites in the world. The site of Dmanisi contains the oldest known human fossils and artifacts outside Africa.

Dmanisi lies at approximately 40°N – about the same latitude as Beijing and Denver. At 1.8 million years ago, when humans are thought first to have occupied the site, climates in the region were only slightly warmer and drier than they are today. The environment has been described as a woodland of pine and birch combined with some open steppic areas. A variety of large mammals roamed this landscape, including deer, horse and other typical Eurasian species.

The human inhabitants of Dmanisi were very similar to the earliest forms of *Homo* in Africa. Their brains were about the same size, and much of the rest of their anatomy was similar as well, although some features of the limb bones suggest that they were better suited to walking long distances. Humans were now living in places where plant and animal foods were less abundant than they are in the equatorial zone. Foraging for food over wider areas was probably one means of coping with these conditions.

Increased consumption of meat was another tactic for living outside the tropics. Meat offers a more efficient way of meeting calorific needs than plants, and it was ultimately essential to human survival in northern latitudes. Mammal bones from Dmanisi exhibit the tell tale cut marks of stone tools. People were either successfully hunting deer and other game or – at the very least – stripping the meat off carcasses before other scavengers could do the same. The tools made and used for these tasks are identical with the African tools associated with the earliest forms of *Homo*.

Artifacts in northern China – at the same latitude as Dmanisi – are dated to between 1.7 and 1.6 million years ago, and reveal that humans were spreading eastwards as well as northwards into Eurasia. The artifacts are similar to those from Dmanisi. The picture in Europe is less clear, but it seems possible if not likely that early humans settled parts of southern Europe at this time. There are sites

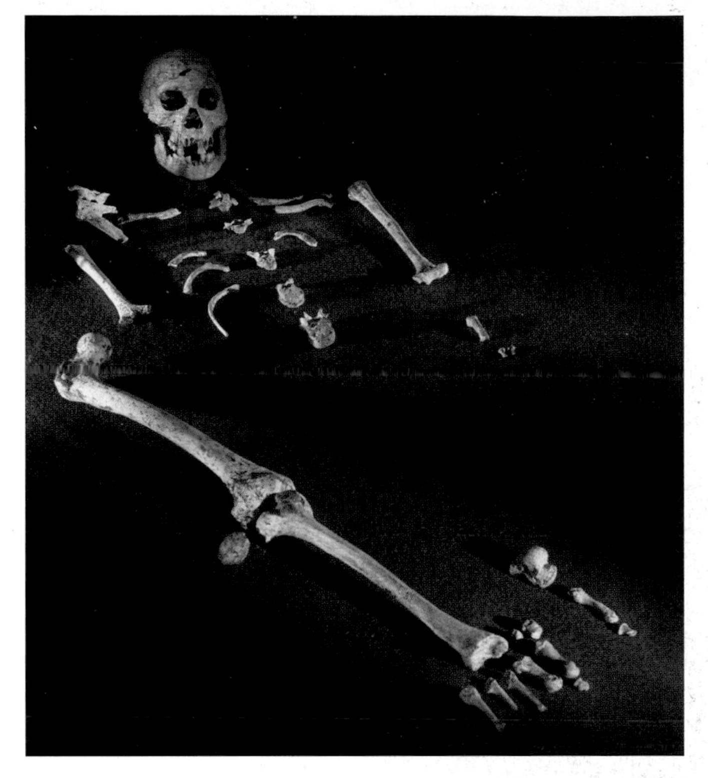

above Human skeletal remains from Dmanisi indicate that the earliest known humans in Eurasia were similar to their African ancestors, although they exhibit some differences in their limb bones that suggest a greater emphasis on walking.

left The stone tools made by the Dmanisi people were also very similar to those of their African ancestors, and probably used for similar purposes, including the stripping of meat off large mammal bones.

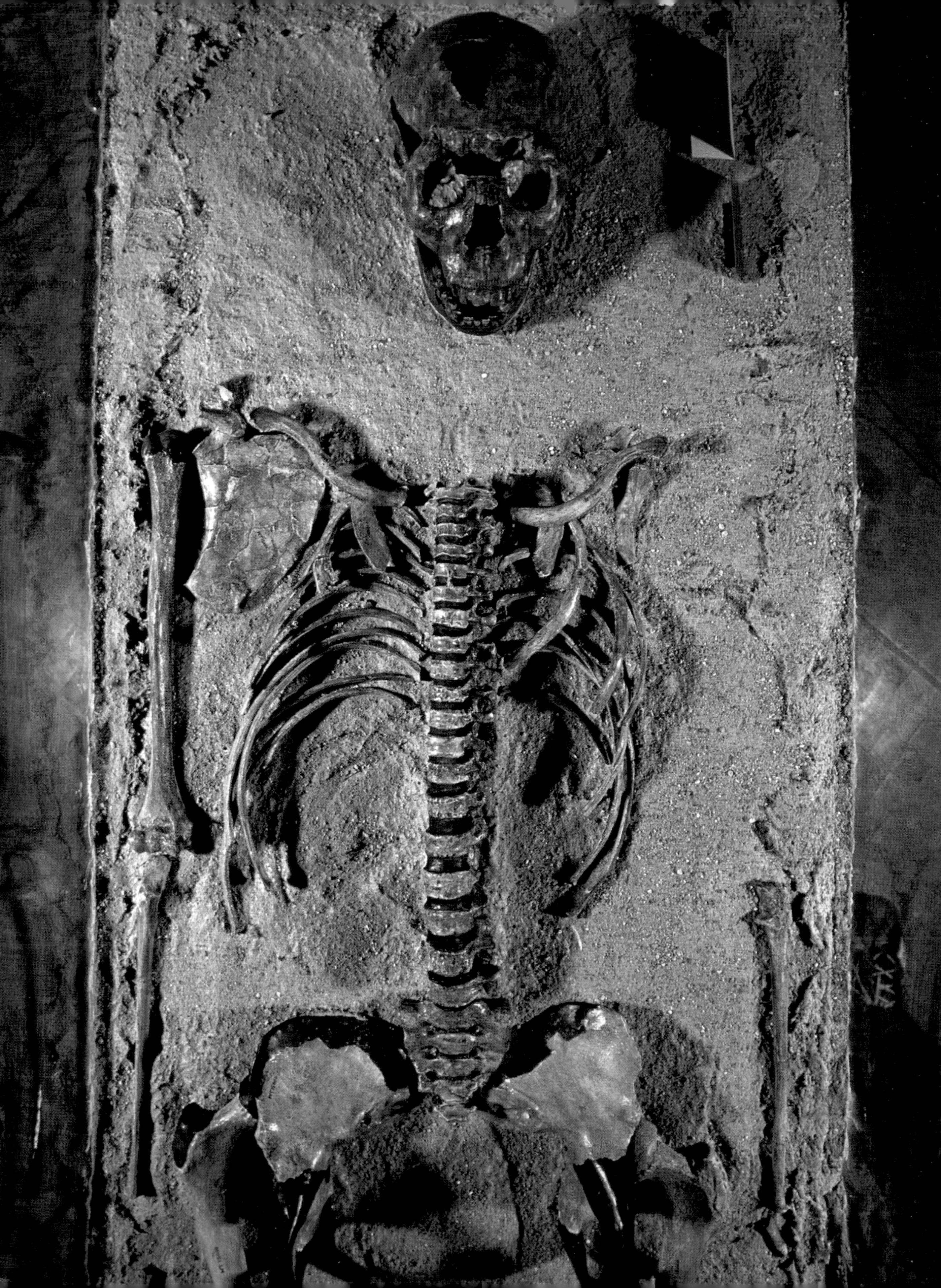

in Spain that may date to more than a million years ago and represent traces of the initial movement out of Africa.

opposite The largely complete skeleton of an adolescent boy recovered from Nariokotome in Kenya and dated to about 1.5 million years ago reveals a person who was more or less modern from the neck down.

Not long after the first expansion of human settlement into Eurasia, some changes took place back in Africa. A remarkable skeleton recovered from Kenya in 1984 – now generally classified as *Homo ergaster* – and dated to about 1.5 million years ago, provides a vivid portrait of an adolescent boy in this setting. Tall and thin – like most of his living descendants of the tropical zone – the youth was essentially a modern human from the neck down, including his hands. Only his skull with its comparatively small brain (about two-thirds the size of ours), heavy brow ridges and large teeth and jaws are strikingly different from our own.

People in Africa began making new types of tools at this time. Between 1.7 and 1.6 million years ago they began to shape large pieces of stone into oval form by chipping them on both sides. Archaeologists refer to this tool as a handaxe (a more pointed form is known as a 'pick' while a blunted version is termed a 'cleaver'). Although the first handaxes were rather crudely rendered, more refined examples soon appeared, and eventually these tools were being made with admirable precision.

Debate over their meaning has continued for centuries. Experiments with handaxes show they make excellent tools for butchering an animal carcass, and microscopic examination of damage along their edges confirms that at least some of them were used for this purpose. But many handaxes were either made and never used or wastefully discarded after limited use. The pattern has invited questions about their economic significance. And throughout much of East Asia, handaxes are rare – most animal carcasses were butchered with other tools.

below Handaxes first appear in Africa between 1.7 and 1.6 million years ago and represent the oldest known example of a mental template imposed upon the external world.

Perhaps the greatest significance of handaxes lies in their implications for human cognition and the ability to articulate thoughts outside the brain – in this case with the hand rather than the vocal tract. Because a finished handaxe bears little resemblance to the original piece of rock from which it was chipped, it represents a concept or mental template imposed upon the world outside the brain. Moreover, the concept was communicated from one individual brain to another and passed down through many generations.

The changes in anatomy and artifacts that took place in Africa at the beginning of the Ice Age soon rippled across Eurasia. The remains of people who were similar in appearance to the gangly youth from Kenya are found in Southeast Asia dating to roughly the same time range. These remains are classified as *Homo erectus*, which is now widely viewed as an Asian variant of the same, or very similar, people that were found in Africa. At a slightly later point in time – about 1.4 million years ago – handaxes appear in the Near East. These probably reflect yet another movement out of Africa.

below A new species of human, often referred to as *Homo heidelbergensis*, evolved in Africa about three quarters of a million years ago and spread north into Eurasia, including western Europe.

opposite below At Boxgrove in southern England, *Homo heidelbergensis* groups moved across a low-lying coastal plain roughly half a million years ago, hunting or scavenging meat from large mammals.

The Heidelberg phenomenon

Roughly three-quarters of a million years ago, a new form of *Homo* evolved in Africa and spread northwards into Eurasia, repeating the earlier pattern. The new people are often referred to as *Homo heidelbergensis* (or Heidelberg Man), which seems odd for a group from sub-Saharan Africa, but reflects the history of fossil discovery. In 1908, a robust human jaw was found in ancient river deposits near the German city of Heidelberg. It now appears that this jaw, along with other skeletal remains in Europe that are about half a million years old, all belong to the same group. These people were more similar to ourselves than any previous form of human. The brain was significantly larger – almost the same size as the modern human brain. While the skull retained some primitive features, it lacked the massive brow ridges and large jaws and teeth of earlier *Homo*.

Homo heidelbergensis continued to make handaxes – as well as cleavers and picks – often with great skill and attention to form. But they also developed a novel way of producing stone flakes for smaller tools that allowed them more control over the size and shape of the flake. The method is usually referred to as the 'prepared-core technique', and it seems eventually to have led to the manufacture of composite tools and weapons (i.e. fitting blades and points into wooden handles or shafts). In addition to this, with *Homo heidelbergensis* we have the first convincing evidence for the use of controlled fire – evidence for this among earlier humans is ambiguous. These improvements in technology – in the ability to manipulate the environment – presumably reflect enhanced cognitive skills, although they seem rather modest when matched against the large increase in brain size that *Homo heidelbergensis* exhibits.

Human fossil remains from this period are scarce in the Near East, and it is the sudden appearance of artifacts similar to those of comparable age in Africa that suggests the arrival of the Heidelbergers. The site of Gesher Benot Ya'aqov in Israel is filled with handaxes and cleavers dating to about 800,000 years ago made from giant flakes of volcanic rock. They are quite similar to the tools that people had been making in Africa from about 1 million years ago onwards. And it is Gesher that yields the oldest traces of controlled fire.

From the Near East or possibly northwest Africa, *Homo heidelbergensis* spread north into Europe, becoming the first human population to establish a substantial presence on the continent and occupy latitudes as far north as the city of London and beyond. Handaxes are found in Italy dating to 640,000 years ago. Roughly half a million years ago – the same age as the Heidelberg jaw – they show up in southern England at the site of Boxgrove.

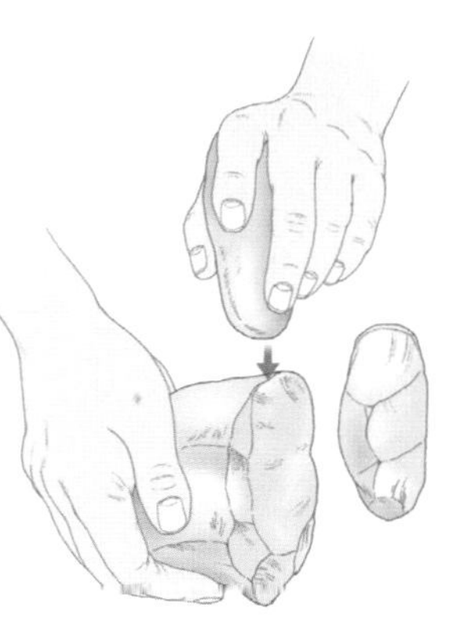

above A new technique of preparing stone cores that allowed greater control of the size and shape of the flakes struck off them was developed roughly half a million years ago.

The Evolution of Human Diet

The ability of humans to obtain and digest a wide assortment of foods has played an important role in their evolution, allowing them to expand their habitat range far beyond their tropical homeland. The variety of foods consumed by living human populations is truly astonishing and exceeds that of all other living organisms. Furthermore, modern humans prepare food in a manner analogous to the use of language – continually recombining elements of animals, plants and inorganic substances into novel recipes, according to cultural tradition and individual taste.

There are two major approaches to the study of past human diet. The more traditional method has been the study of food remains found in association with human fossils and artifacts. Such remains are typically the bones and teeth of animals recovered from archaeological sites, but plant remains also have been retrieved – especially in recent years – from Ice Age sites. Because animal and plant remains can be introduced into archaeological sites by various non-human agencies – carried into a cave, for example, by its non-human residents or washed into an open-air occupation by running water – the methods of *taphonomy* (the study of processes that affect an organism after death) are employed to determine their origin. Animal bones accumulated by a hyaena, for example, will exhibit certain identifiable characteristics with respect to the parts of the skeleton represented, types of damage to bone surfaces observed, and others.

More recently, the chemical analysis of human bone and teeth has provided information on the contribution of different types of food to the diet. Many foods contain varying amounts of stable isotopes of carbon, nitrogen and other elements. These isotopes become incorporated into the chemistry of the bone and since they do not decay over time they can be measured long after the death of the individual who consumed the food.

left Cut marks on bones from Boxgrove confirm that humans there were stripping meat off large mammal carcasses half a million years ago.

opposite As humans moved northwards, protein and fat from animals such as bison became increasingly important in their diet.

Humans inherited an omnivorous diet from their primate ancestors; living apes consume a relatively eclectic diet of leaves, fruits, insects and other plant and animal foods of the tropical forest and woodland. But the analysis of australopithecine fossils reveals an early expansion of diet in Africa to include either plant foods of the savannah (sedges and grasses) or the animals that ate these foods, or both. The oldest known archaeological sites – occupied by early *Homo* and possibly one of the robust australopithecines – contain large mammal bones that have been broken and scraped with stone tools.

As humans expanded out of the tropical zone and into the cooler and less productive environments of temperate Eurasia during the Ice Age, their consumption of meat almost certainly increased. It has been difficult to determine exactly how much meat they were eating and how they came by it – how much of it was scavenged rather than hunted. Taphonomic study of large mammal bones from sites occupied by *Homo heidelbergensis* and its contemporaries between roughly 750,000 and 250,000 years ago confirms that meat was being removed from carcasses, but cannot answer the question of how many of the carcasses represent hunted prey or what percentage of the diet they provided.

By contrast, the later Neanderthals present a clear pattern of large-mammal hunting and heavy consumption of meat. Stable isotope analyses of Neanderthal bone reveal that most of their protein was derived from animal sources – even during intervals of warm climate when digestible plant foods must have been plentiful. Taphonomic studies confirm intense processing of carcasses and bones, although they also show that Neanderthals often shared their camps with various carnivorous competitors. A diet rich in protein and fat was probably a critical part of how Neanderthals coped with cold Ice Age environments in northern Eurasia.

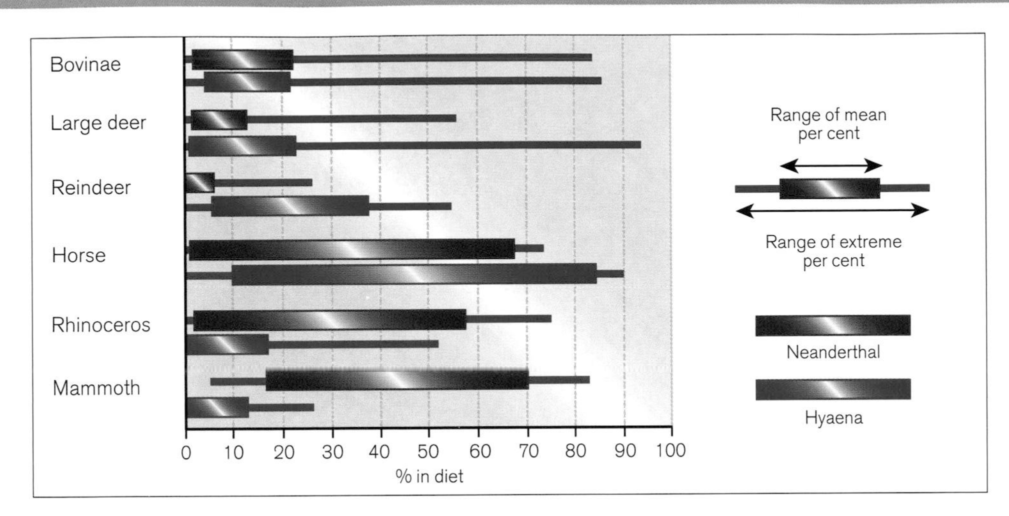

As shown in this graph, chemical analysis of a Neanderthal bone from western Europe revealed an an emphasis on consumption of the largest land mammals available – mammoth and rhinoceros. These animals would have provided a rich source of protein and fat, and hunting them would have reduced food competition with the Neanderthals' ever-present rivals, hyaenas, which usually consumed smaller mammals.

The appearance of modern humans in these environments after 50,000 years ago marks a significant and revealing shift in diet towards a much wider range of animal – and probably plant – foods in comparison to the Neanderthals. Modern humans broadened their range of harvested prey to include small mammals, birds and fish. The shift is documented by both the animal remains found in their sites and the stable isotope analysis of human bone. There is also growing evidence for the use of plant foods, in north Eurasian sites occupied by modern humans during the last glacial period. Their ability to create novel and often complex technologies (e.g. throwing darts, nets, boats and weirs) undoubtedly was critical to this development and it underlies the subsequent shift to agriculture in some parts of the world – and shift to a heavy marine economy in other places – following the end of the Ice Age.

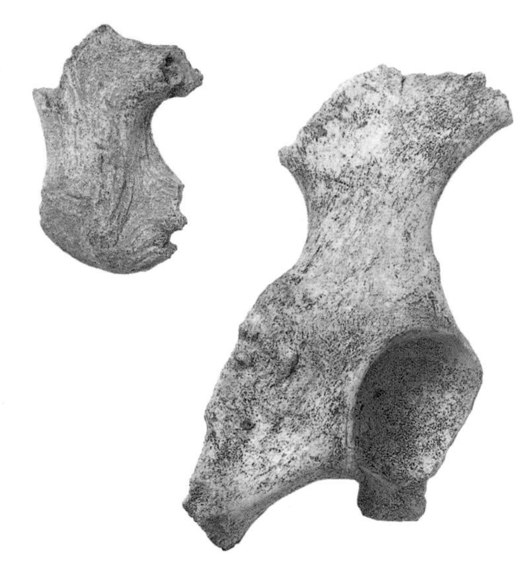

left In addition to stone artifacts, the occupants of Boxgrove manufactured tools from bone and antler. On the left is a soft hammer made from antler for stone working; on the right is a butchered pelvis, probably from a rhino.

right Exceptional preservation conditions at the 400,000-year-old site of Schöningen in northern Germany yielded long wooden spears or probes with sharpened points.

Carefully excavated by archaeologists since the early 1980s, Boxgrove represents more than a site. It is actually an ancient landscape. Then, as now, it was part of the low-lying coastal plain, below eroding limestone cliffs. Climates were warmer then, and horse, rhinoceros and other large mammals roamed across a grassy plain. Humans moved among them, hunting or scavenging meat. They gathered repeatedly around what was probably a spring or waterhole near the base of the cliff, where they often discarded lightly used handaxes. At other locations, they sharpened their tools and used them to butcher the carcass of a large mammal.

Another extraordinary site – somewhat younger, and occupied under cooler conditions – is found at Schöningen in northern Germany. Roughly 400,000 years ago, the area was covered with meadows and forest steppe. As at Boxgrove, horses were butchered here; the bones were smashed and scraped with stone tools. But unlike Boxgrove, Schöningen yielded remains of former fireplaces. It also produced several wooden artifacts: spears and possibly other implements. Analysis of the spears reveals a rather complicated sequence of manufacture. Each spear was made from a small pine or spruce tree. The base of the tree – where the wood is hardest – was used for the point, which was shaped with great care, then polished and cleaned.

What is unclear is the extent to which the Heidelberg people had become adapted to cold climates. By half a million years ago, Europe was well into the pattern of glacial cycles that saw an alternation between extreme cold and intervals of warmth similar to those of the present day. Although most sites were occupied during the warm interglacial periods, some – like Schöningen – were visited during cooler times, and a few sites contain evidence of a human presence during the glacial periods. One of these is Boxgrove, which yielded some artifacts in younger glacial sediments that overlie the main occupation level.

Because of the relative scarcity of human fossils – especially bones of the post-cranial skeleton (i.e. below the neck) – it is difficult to determine if *Homo heidelbergensis* had evolved some of the anatomical features that we expect to see among people living in very cold places. In order to prevent loss of body heat and protect against frostbite, people (as well as other warm-blooded animals) tend to evolve shorter limbs and a heavier body mass. A lower leg bone recovered from the main occupation level at Boxgrove is long – comparable to that found with the 1.5-million-year-old skeleton from Kenya mentioned earlier – and better suited to life in the tropics. On the other hand, the leg bone is very robust, indicating that it belonged to someone (probably an adult male) with a large body mass. It is only a single bone, however, and may not be representative of the whole *Homo heidelbergensis* population.

Increased meat consumption – a diet higher in animal protein and fat – is another strategy for survival in cold environments, as already noted. But here again the evidence is ambiguous. While Boxgrove, Schöningen and other sites provide ample evidence for stripping meat off large mammal bones, the quantity of meat consumed may not have been significantly higher than that consumed in earlier times and more southerly latitudes.

Perhaps these people relied on new forms of technology. Control of fire, which is currently linked to the appearance and spread of *Homo heidelbergensis*, is the most obvious of these. Not only would fire have provided much-needed warmth in freezing conditions, but it would have also allowed people to cook and digest the higher-calorie food needed for survival in a cold climate. In addition to this, fire would have given welcome protection from the formidable beasts that inhabited the new areas of the world that *Homo heidelbergensis* was colonizing. More efficient tools and weapons – exemplified by the well-crafted wooden spears from Schöningen – might have enhanced their ability to acquire food in a less productive environment.

A human tibia or lower leg bone from Boxgrove indicates a relatively robust individual with long legs. The latter suggests that *Homo heidelbergensis* did not evolve some of the anatomical adaptations to cold climates, such as a short, stocky build, found among later human occupants of northern latitudes.

In East Asia, the human response to repeated intervals of intense cold during this period is less hard to read. During the glacial episodes, the Asian contemporaries of *Homo heidelbergensis* simply abandoned the northern areas of the continent. At the cave of Longgushan in northern China, traces of human occupation are found only in layers that date to the intervals of warm climate between 740,000 and 400,000 years ago. During the glacial intervals, they apparently retreated southwards to the subtropical zone.

Even during the warmest periods, these people were living no farther north than Dmanisi and other places occupied more than 1.5 million years ago by earliest *Homo* (i.e. about 40° north). The contrast with western Europe – where *Homo heidelbergensis* was living as far as 52°N – probably reflects the mild local climates created by warm currents in the Atlantic Ocean.

It might also reflect some differences in the people who inhabited the Far East at this time. Unlike *Homo heidelbergensis*, they were not recent arrivals from Africa, but rather descendants of an earlier movement into Eurasia. They are classified as *Homo erectus* – essentially the same as the earliest known humans of East Asia. For the most part, they were still making pebble choppers and other stone tools similar to

those made in Africa more than 2 million years ago. Like the elegant handaxes of *Homo heidelbergensis*, they were being used to butcher large mammal carcasses, but their simplicity may reflect a generally low ability to manipulate the world outside the brain – to create complex technology. On the other hand, Longgushan Cave contains burnt animal bone that seems to confirm control of fire.

In July 1997, a team of laboratory scientists published the results of a remarkable and successful effort to extract and analyse DNA from the bone of a direct descendant

Tangshan Cave (also known as Nanjing), 250 km (155 miles) northwest of Shanghai in eastern China, has yielded fossils of *Homo erectus* dating to roughly 600,000 years ago.

Fire and Human Adaptation: The Archaeological Evidence

The control of fire has long been regarded as a fundamental human invention that set humans apart from other animals at an early point in their career. Controlled fire has also been viewed as critical to the conquest of Ice Age environments in the Northern Hemisphere. Documenting the presence of fire in the archaeological record has been difficult, however, because it is surprisingly hard to see in that record. Former hearths do not preserve over long periods of time, while traces of natural wildfires often mimic the effects of controlled fire.

It is entirely possible that only modern humans mastered the technology of generating fire and that Neanderthals and earlier humans were confined to simply controlling it – having obtained it from a natural source. Among recent foraging peoples of the tropical and temperate zones, several reportedly lacked the ability to produce fire, which underscores the complexity of this technology as well as the possibility of living without it. Nevertheless, from *Homo heidelbergensis* onwards (i.e. about three-quarters of a million years ago) humans at least possessed the knowledge and skill to manage fire and use it in their campsites, as well as perhaps other settings.

The earliest reported evidence of controlled fire consists of burned clay in African sites that are more than 2 million years old. The evidence is ambiguous, however, and the burned clay patches may simply represent traces of wildfire. More plausible evidence dates from 1.5 million years ago in South Africa. Levels dating to this time range at Swartkrans Cave – and occupied by *Homo ergaster* – contain burnt bone that is lacking in the underlying layers. While bone could also be burnt by natural fire, its sudden appearance in the upper levels of the cave seems odd and is perhaps more easily explained by human action.

For many decades, the cave of Zhoukoudian in northern China was said to contain the oldest traces of hearths, dating to roughly half a million years ago. The conclusion was based primarily on the discovery of black layers that appeared to be ash and charcoal. Re-study of these layers a few years ago determined that they were decomposed plant remains deposited by flowing water. The cave also contains burnt bone, however, which – as at Swartkrans – may represent controlled fire.

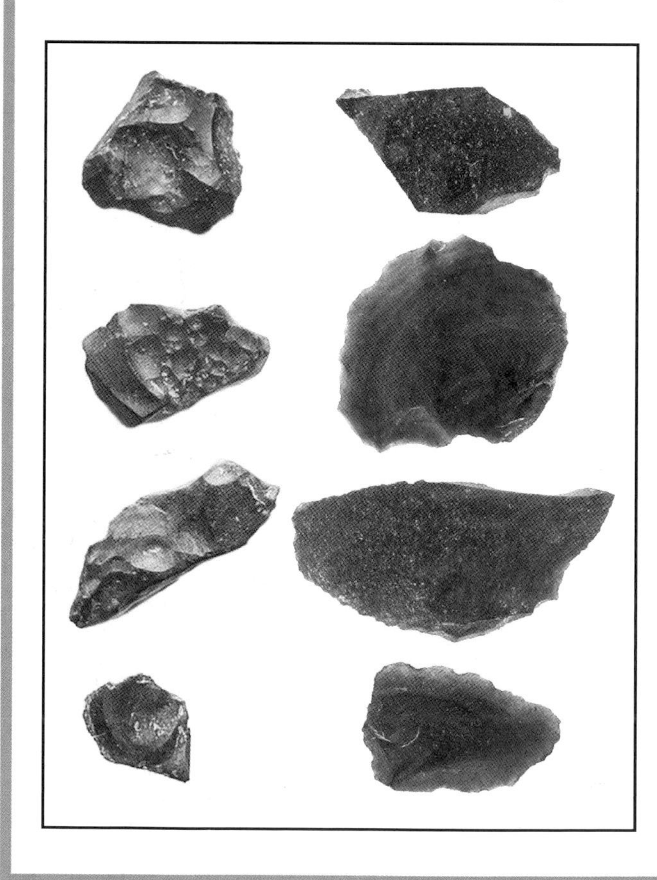

left Pieces of burnt flint, wood and seeds found at the site of Gesher Benot Ya'aqov in Israel and dated to almost 800,000 years ago indicate controlled use of fire by *Homo heidelbergensis*.

New and convincing evidence has recently been reported from an even older context in the Near East. At Gesher Benot Ya'aqov in Israel, concentrations of burned flint, wood and seeds dating to almost 800,000 years ago probably reflect the intense and localized heat generated by burning hearths. This site was apparently visited by *Homo heidelbergensis* prior to its appearance in Europe. Several major *Homo heidelbergensis* sites in Europe also contain reliable evidence for controlled fire, including Schöningen (Germany) dating to about 400,000 years ago.

Former hearths are common in Neanderthal sites – caves and rockshelters as well as open-air sites – dating to less than 250,000 years ago. In part, the pattern reflects the comparatively young age of these sites and the fact that many are found in natural shelters. But it probably also reflects intense use of controlled fire in the glacial landscapes of northern Eurasia. Modern humans, who invaded these landscapes after 50,000 years ago, continued to make intensive use of fire for warmth, protection and food preparation, but they also began to develop various technological applications of controlled fire, including fired ceramic technology. They also expanded the range of fuel to include bone, coal and animal fat (used in portable lamps).

below Traces of former hearths are common in Neanderthal sites and indicate a heavy use of controlled fire. It is not known if the Neanderthals had mastered the technology of making fire, however, and they may simply have scavenged it from natural wildfires.

of *Homo heidelbergensis* – a Neanderthal (named after the Neander Valley in Germany, where the first remains were found). Since then, DNA has been recovered from other Neanderthal specimens excavated from archaeological sites in various parts of Europe. The principal objective of these studies has been to assess the genetic relationship between the Neanderthals and ourselves. The genetic distance between the two species suggests a split in our common lineage roughly 600,000 years ago. When matched up with the fossil record, Neanderthal DNA indicates that both modern humans and Neanderthals are descended from *Homo heidelbergensis*, which – by expanding out of Africa at roughly this time – gave rise to northern and southern forms of human. We are derived from the southern or African form, while the European descendants of *Homo heidelbergensis* evolved into the quintessential Ice Age people – *Homo neanderthalensis*.

The Neanderthal alternative

The Neanderthals are at once the best-known and most mysterious of all early forms of human. They are best-known because of the great wealth of fossil material recovered from Europe and the Near East over the past 150 years, as well as the

Sites occupied by the Neanderthals are found all across Europe and extend into the Near East and Central Asia. The Neanderthals probably occupied the Altai region of southwestern Siberia as well, although only isolated fragments of their presumed skeletal remains are currently known from this region.

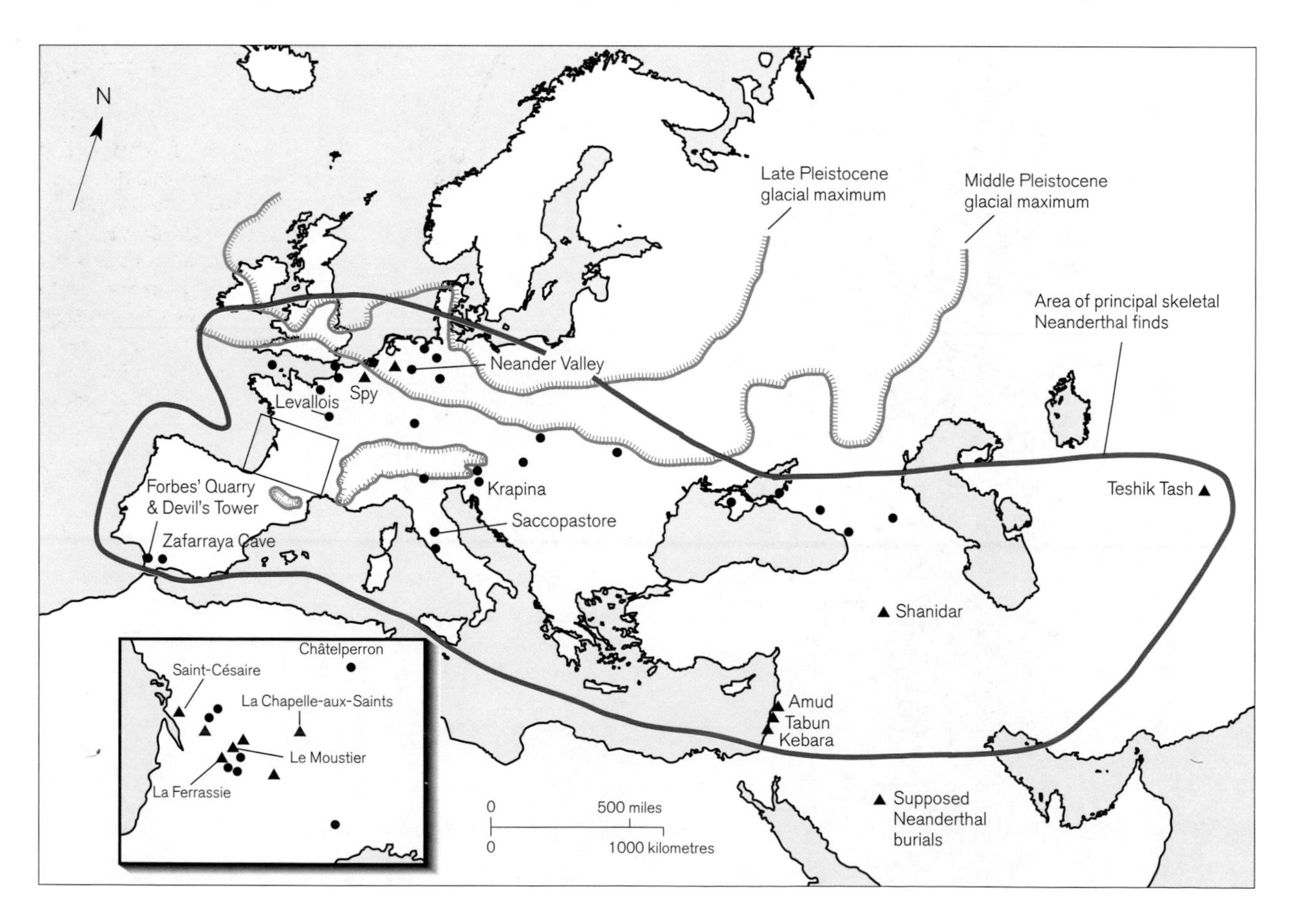

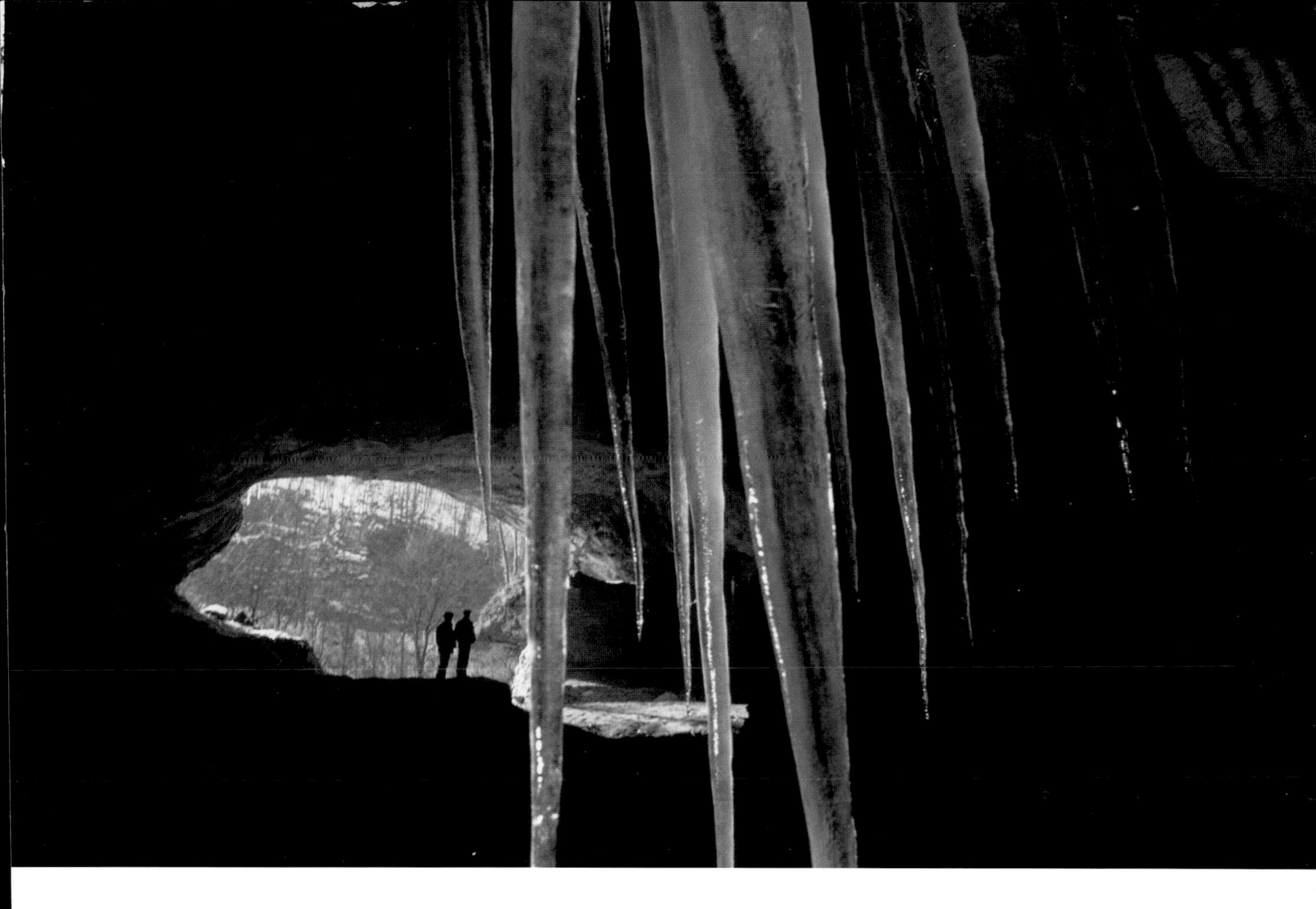

considerable number of Neanderthal archaeological sites excavated over the same period. This reflects in part their comparatively recent age (they survived as late as 30,000 years ago) and an abundance of caves and rockshelters, where their remains are well preserved and easily found. They are mysterious because they behaved like modern humans in some ways – they buried their dead, for example – but were very different from us in other ways. Given our common recent ancestry in *Homo heidelbergensis*, the Neanderthals could be considered an *alternative* form of ourselves, rather than another early human ancestor.

The Neanderthals were a northern form of human in the same way that the Arctic hare is a northern variant of the jackrabbit, and they are the only known such form of human. Like other warm-blooded mammals living in cold places, they evolved a suite of anatomical features that helped them endure extreme low winter temperatures. They were stocky and barrel-chested with shortened forelimbs, all of which served to conserve body heat and minimize the danger of frostbite. Below their eye sockets, unusually wide openings in the cheek bone (*suborbital foramina*) allowed a generous supply of blood to the face. The braincase was large, and while this too might have been an adaptation to cold, its climatic significance may have been outweighed by the value of a large brain.

Neanderthal skeletal remains and associated artifacts are dated to less than 40,000 years ago at Vindija Cave in Croatia. Much of the reconstructed Neanderthal genome – the 'first draft' of which was announced in February 2009 by the Swedish geneticist Svante Pääbo – is based on fossil DNA extracted from these skeletal remains.

The Neanderthals evolved gradually in western Europe from the local *Homo heidelbergensis* population. By 300,000 years ago, many of their characteristic traits can be seen in the European fossil record. Besides their anatomical adaptations to cold, the Neanderthals developed other traits that set them apart from modern humans (also evolving from a local *Homo heidelbergensis* population in Africa at this time). The top of the Neanderthal head was low and flat with heavy brow ridges that extended over the eyes like early forms of *Homo*. The nose was large, with massive nasal cavities, and the mouth projected so far forward that a gap existed between the third molar and the ascending ramus of the jaw.

above Personal ornaments in the form of perforated animal teeth are widely thought to have been produced by Neanderthals in western Europe during the final millennia preceding their disappearance.

While uncertainties remain about the ability of their European predecessors to cope with Ice Age glacial periods, the Neanderthals are closely linked to cold environments. In contrast to the pattern at Longgushan Cave, Neanderthal occupations in western Europe span the most severe glacial intervals without interruption. In caves and rockshelters of France and Spain, their artifacts and other camping debris are found in layers filled with angular fragments of rock that were spalled off the walls and ceiling by the effects of extreme cold. These dwellings also contain the bones of reindeer and other residents of the modern circumpolar zone.

The Neanderthals expanded eastwards into colder regions of northern Eurasia that no earlier human had ever inhabited. Today, the average January temperature in the city of Paris is roughly 3°C (37°F). But if you walk east towards Russia, the climate becomes drier and the winters become colder due to the waning influence of the North Atlantic. By the time you reach the city of Volgograd – at the same latitude as Paris – the average January temperature has dropped to a frigid –7°C (19°F). During the glacial periods of the Ice Age, temperatures fell all across northern Eurasia, and the January mean for the Volgograd region would have been very low (probably about –20°C, or –4°F).

The Neanderthals became the first humans to settle widely across the vast plain of eastern Europe, which stretches over 1,600 km (1,000 miles) from the Carpathians to the distant Ural Mountains. Earlier human occupation had been confined to places along the southern fringe of this plain. Some of the Neanderthal sites on the central and southern parts of the plain date to a warm interglacial period roughly 125,000 years ago that preceded the final glacial period of the Ice Age. But most of their sites date to intervals when climates were colder than those of the present day. During the earlier phases of the last glacial period, they occupied sites along major river valleys in the southwestern part of the plain – in what is now southwestern Ukraine.

At sites like Molodova in the Dnestr River Valley, winters must have been bitterly cold – with brisk winds from the eastern slopes of the Carpathian Mountains blowing down the valley. Moreover, if the transportation of rock from its source area is any guide to the movement of the people who carry it, the Neanderthals operated within relatively small territories. The stone used for tool production in their sites is usually derived from a source less than 80 km (50 miles) away (in very rare cases, stone was moved as much as 220 km, or 135 miles). In short, the Neanderthals were probably year-round residents of these areas – not summer visitors from warmer regions in the south.

The Neanderthals seem to have been the first humans to inhabit some of the colder regions of northern Eurasia, including the central plain of eastern Europe, where winter temperatures are significantly lower than those of western Europe.

Nor did the Neanderthals confine their eastward expansion to eastern Europe. They subsequently occupied the southwestern portion of Siberia – where climates are even drier and colder. Their artifacts and at least a few isolated skeletal remains are found in caves of the mountainous Altai region (several thousand feet above sea level). Their way of life seems to have been similar to that of Neanderthals in Europe, but the Altai people would seem to have lived under the harshest conditions in the Neanderthal world.

A recent study concluded that the Neanderthals probably relied less on their anatomical adaptations to low temperature than previously believed. Their ability to produce effective technology and procure a diet rich in animal protein and fat was equally – if not more – important for Ice Age cold weather survival. Some new research reveals significant advances over earlier humans with respect to the design of weapons and tools. The Neanderthals were making composite implements: weapons and tools assembled from several parts. Stone points were attached to shafts to produce deadly thrusting spears, while chipped stone flakes were attached to wooden handles for cutting and scraping tools. The leverage provided by the handles would have considerably improved the efficiency of such tools over hand-held pieces of stone.

The Neanderthals produced blanks from stone cores, and fashioned composite weapons and tools by attaching these blanks to wooden shafts and handles.

Although not a single composite implement has been preserved – wood rarely survives in Ice Age sites – their presence is indicated by at least two lines of evidence. Microscopic analysis of stone artifacts from Neanderthal sites reveals subtle traces of wear caused by friction from a wooden handle or shaft. The surfaces of some stone points and flakes also exhibit faint residues of the adhesives used to attach them to the latter. In one case, the adhesive was pine resin, and in another site it was bitumen.

Less information is available about Neanderthal clothing. Traces of polish on some stone artifacts confirms that the Neanderthals were scraping hides, and it is difficult to conceive of how they might have endured winters in places like the Dnestr Valley and the Altai Mountains without some form of protective clothing. Nevertheless, their sites are completely devoid of a small and simple tool – the sewing needle – that is essential to the production of the sort of clothing created by later peoples of the Arctic. Neanderthal clothing was probably less complicated – and consequently less effective – than the tightly-sewn tailored fur clothing found among the arctic peoples. Perhaps this was a critical limitation on their geographic range. The lack of sewn winter clothing may be the reason why the Neanderthals avoided the most extreme climates in northern Eurasia, which are found in the interior parts of northeast Asia.

Throughout their range, the Neanderthals were accomplished hunters of large mammals. Analysis of their bone chemistry shows that their protein was almost entirely derived from animals, and a diet rich in animal protein and fat was undoubtedly essential to their existence, providing for their high calorific needs in a cold-climate setting. In the caves of southwestern France, Neanderthals feasted on deer meat, smashing the bones into pieces to retrieve the marrow. Further east – where the landscapes were drier and the flora more steppic in character – they hunted bison and saiga (a goat-like antelope) in large numbers. In the mountains of the Caucasus, they hunted goats and sheep.

At La Cotte de Saint Brelade on the island of Jersey in the English Channel, Neanderthal hunters apparently drove mammoth and woolly rhinoceros over a cliff.

A recent study of Neanderthal bone chemistry in France yielded a surprise. Employing new techniques that permit the analyst to identify specific animals that contributed to the diet, the study revealed that mammoth and woolly rhinoceros had been a major source of protein. The bones of these massive herbivores are rare or absent in the caves of southwest France and some archaeologists have wondered whether the Neanderthals were capable of hunting them. But their absence may simply reflect a lack of enthusiasm for dragging the heavy bones back to camp – most large animals were probably killed some distance from the caves where the meat was stripped off the bones.

At one of the few open-air sites in this part of Europe, La Cotte de St Brelade, which is located on what is now one of the Channel Islands but was then joined to the mainland, archaeologists recovered a large number of mammoth and rhinoceros remains – apparently animals that had been driven over a cliff by Neanderthal hunters. And in eastern Europe, where caves are less common and most sites are open-air locations, mammoth bones are sometimes found in great numbers

left The European Neanderthals were barrel-chested with large heads and shortened forelimbs that probably reflect anatomical adaptation to cold climates.

(for example, at Molodova in the Dnestr Valley). These sites seem to confirm the bone chemistry analysis, supporting the notion that Neanderthals were hunting the largest mammals of Ice Age Europe on a regular basis.

The spectacle of Neanderthals hunting and butchering mammoths – which seems almost certain to have been done by groups – raises questions about Neanderthal society. How large were Neanderthal groups and how were they organized? In the few places where the boundaries of a Neanderthal settlement can be defined, the area is small and the inference is that the group was also small – perhaps no more than a dozen people. A small group size would be consistent with the observation (noted above) that Neanderthals probably moved around in relatively small territories. Reconstructing the organizational structure of Neanderthal society is almost impossible. Perhaps it was based more heavily on blood relationships than a modern human society – a pair of siblings with their mates and offspring, for example.

The absence of symbolism and art in their sites – sculptures, engravings, cave paintings and so forth – also raises questions about Neanderthal language. Many archaeologists have speculated that the Neanderthals lacked a language comparable to our own – a syntactical language with phrase structure grammar that would have permitted an infinite variety of sentences and narratives. There have been several attempts to reconstruct the vocal tract and other aspects of Neanderthal anatomy related to speech, but the results have been problematic and the issue remains unresolved. In general, Neanderthal material culture seems to lack the unrestrained creativity that we see in syntactical language. By contrast, all aspects of modern human culture – including our artifacts – exhibit creativity and constant change and diversification.

In one respect, however, the Neanderthals share a unique practice with modern humans. The Neanderthals buried their dead – at least on occasion –

below A Neanderthal skeleton from Kebara Cave in Israel. Neanderthals who lived in the Near East under a somewhat milder climate regime exhibit less extreme anatomical adaptations to cold.

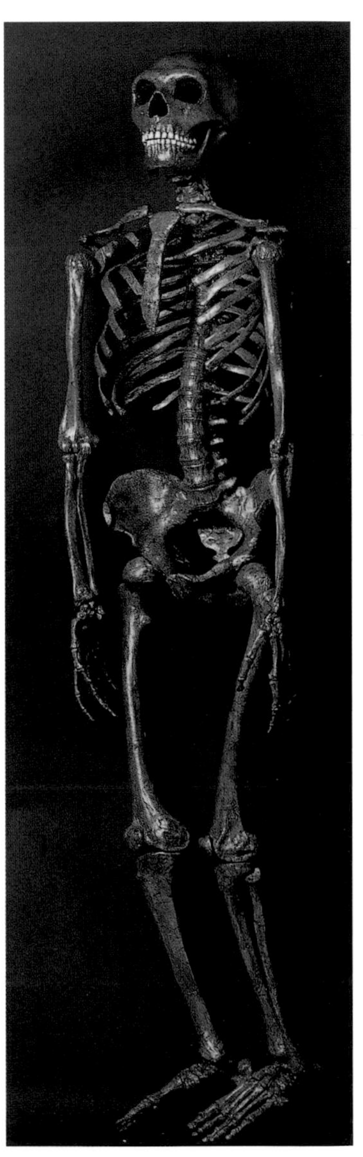

and Neanderthal graves have been found in caves and rockshelters from Spain to Uzbekistan. Some archaeologists have questioned these findings, but the consensus falls on the side of intentional burial. On the other hand, evidence of funerary ritual associated with these burials is ambiguous and inconclusive. Sometimes, artifacts and bone fragments found in the burial pit have been interpreted as funerary objects or 'grave goods' interred with the deceased. But no one can be certain they were not inadvertently shovelled into the pit – the burials are found in habitation sites littered with the debris of Neanderthal occupation.

Why did the Neanderthals bury their dead? A few archaeologists suggest it might simply reflect disposal of waste, but most suspect that deeper motives lie behind this practice. At the very least, it may indicate an awareness of the inevitability of death – a psychological burden that among living organisms only humans bear today.

The emergence and spread of modern humans

The last known Neanderthal was laid to rest in a cave in southern Spain about 30,000 years ago. From that point onwards, modern humans ruled Europe, along with the rest of the world. Despite more than a quarter of a million years of success in the Ice Age environments of Europe – as well as expansion into some of the colder parts of Eurasia – the Neanderthals were replaced or absorbed by people from the tropics. And because the latter were singularly ill-suited for life in Ice Age Europe and Siberia, the transition invites scrutiny. The degree to which the Neanderthals contributed to what followed – either genetically or culturally or both – is a major point of contention in palaeoanthropology.

Modern humans or *Homo sapiens* were the southern descendants of *Homo heidelbergensis*. They evolved gradually in Africa during the same period that the Neanderthals were evolving their characteristic features in Europe and they subsequently expanded into Eurasia just as earlier forms of *Homo* had spread out of Africa. This time, however, the expansion was different. Modern humans spread – rapidly it seems – into an astonishing range of habitats and climate zones without differentiating into new species. They were the first primates of any sort to advance into the Arctic – at least on a seasonal basis – and they also invaded desert margins, boreal woodland, rainforest and cold wooded steppe. Modern humans eventually inhabited the most extreme Ice Age environments of northern Eurasia, including places that even the cold-adapted Neanderthals had been unable to conquer. They accomplished this with an unprecedented capacity for creativity and innovation – for adapting to environmental challenges through novel behaviour and invention. This capacity reflects the emergence of the mind.

Modern humans in northern Eurasia invented the eyed needle – and by implication sewn clothing – no later than 35,000 years ago, which reflects their ability to design new and complex technology to adapt to varying environmental conditions.

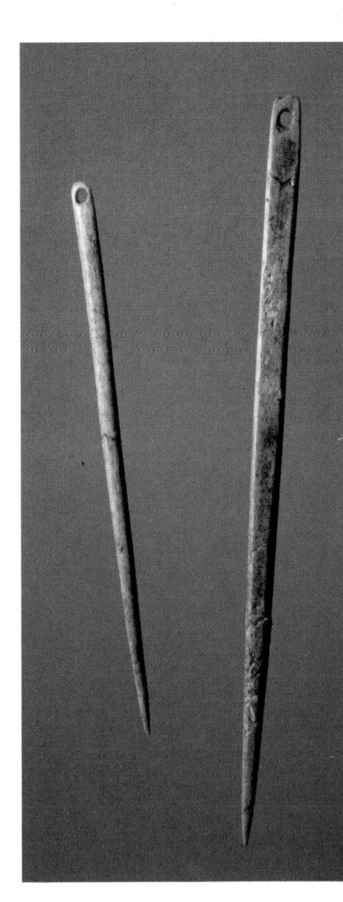

Modern humans spread out of Africa no later than 50,000 years ago, rapidly occupying a variety of habitats and climate zones across Eurasia and Australia. Evidence for an earlier presence in southern China remains problematic, while modern human remains in the Near East dating to roughly 100,000 years ago seem to be unrelated to the main dispersal event.

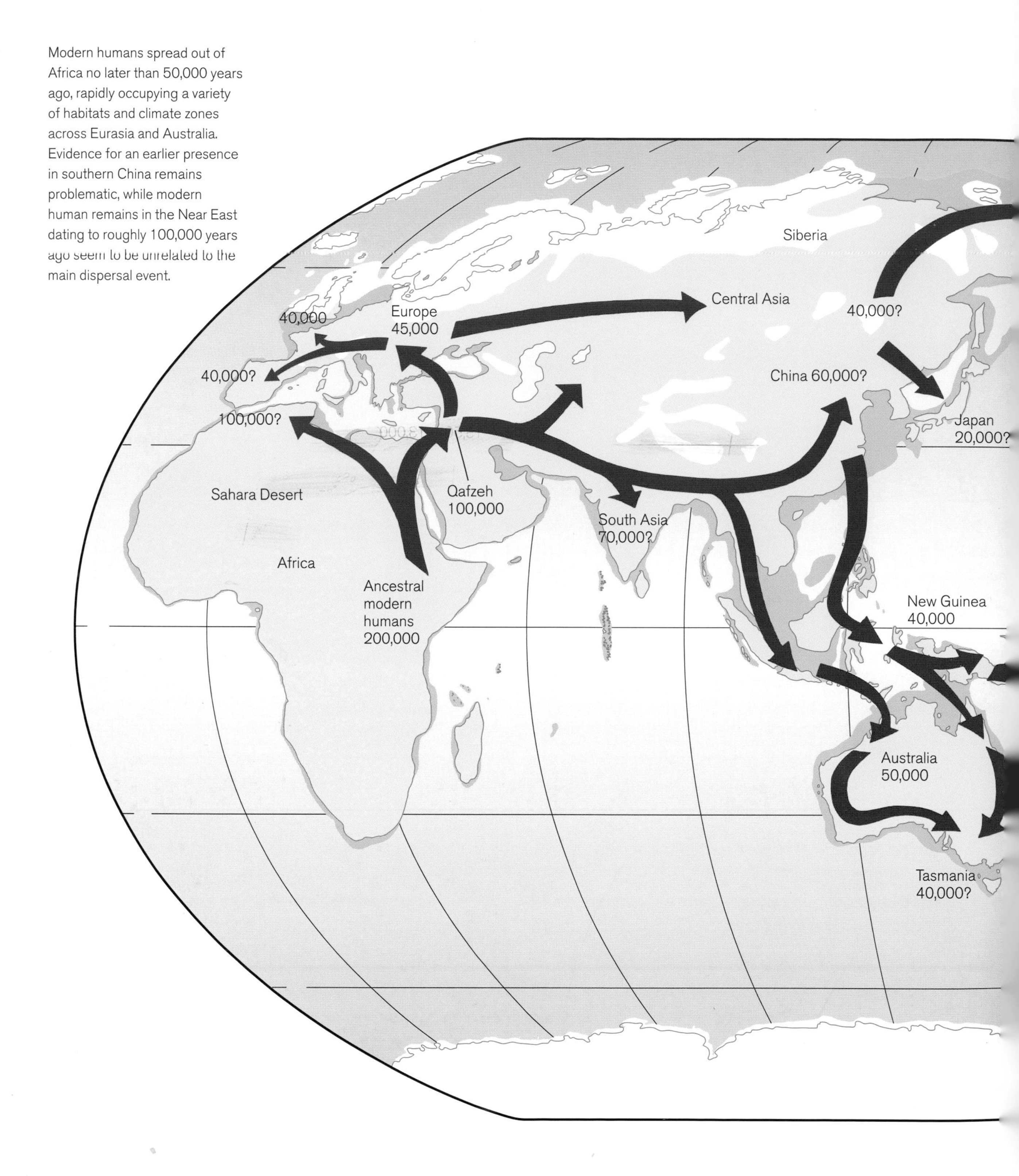

By 30,000 years ago, modern humans were present above the Arctic Circle in northeast Asia, but they do not seem to have settled the Americas in substantial numbers until the final millennia of the Ice Age. The occupation of Oceania was completed only a few thousand years ago.

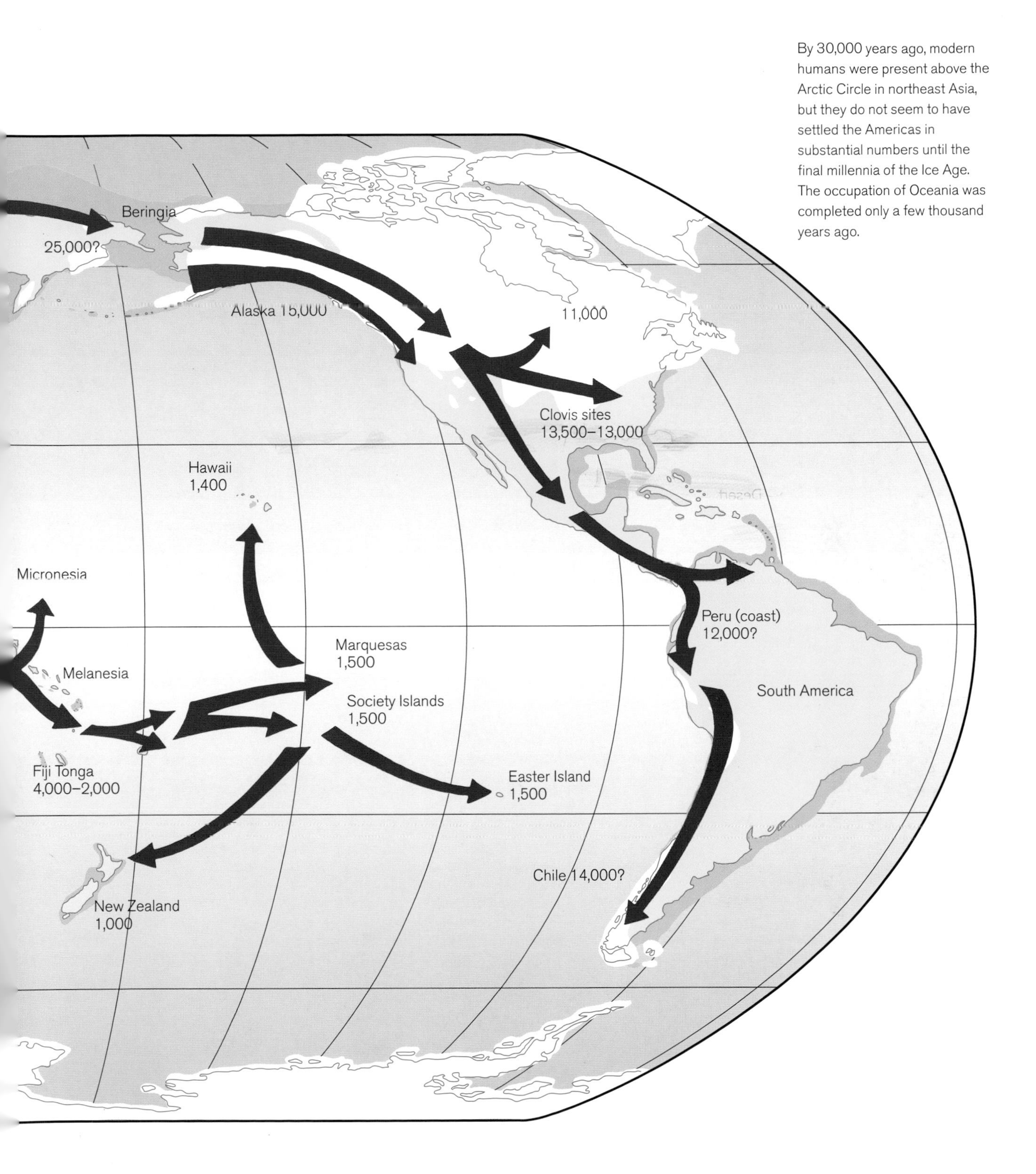

There may be something of a disconnect between the anatomy of modern humans and archaeological evidence for what palaeoanthropologists recognize as 'modern behaviour' (i.e. evidence of the mind). Fossil remains of anatomically modern humans are present in Africa by 200,000 years ago – a date obtained from a skull fragment from Ethiopia several years ago. In fact, there are even older remains from east and South Africa that suggest modern humans – or people very close to us in physical appearance – had evolved at some point after 400,000–300,000 years ago.

Many anthropologists view language – specifically syntactical language – and the use of symbols as the essence of modern human behaviour; language may be the key to how the mind creates. But spoken words cannot be preserved in the archaeological record and the presence of language prior to the invention of writing must be inferred from other sources. A small stone engraved with a series of simple geometric patterns recovered from Blombos Cave in South Africa and dated to about 75,000 years ago may represent the earliest known example of art or abstract symbolic expression. Blombos Cave also yielded some points and awls shaped from bone that, while comparatively simple in design, nevertheless reflect some new developments in technology.

Blombos Cave in South Africa yielded the earliest known example of abstract design – a small block of red ochre incised with a series of simple geometric patterns dating to about 75,000 years ago.

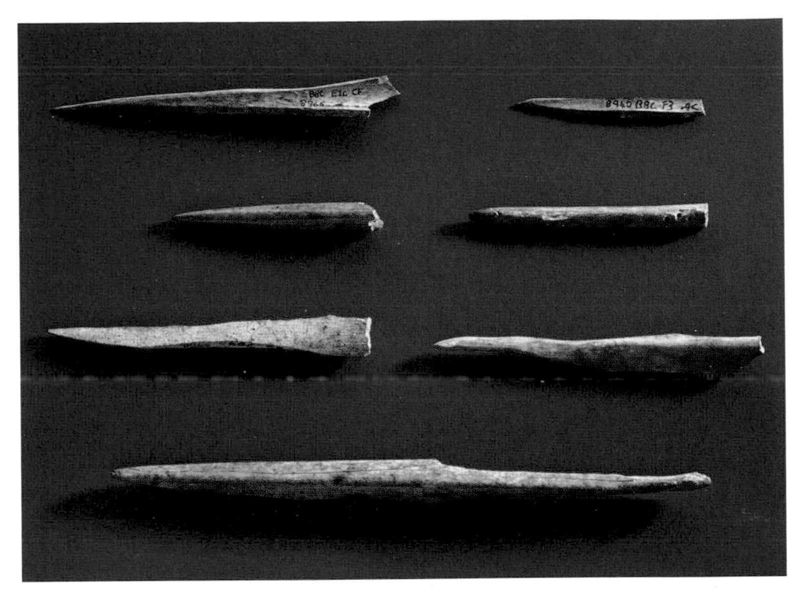

At Blombos Cave in South Africa, modern humans made bone awls roughly 75,000 years ago. These awls may have been used for piercing and perhaps engraving items.

From older sites in Africa – more than 200,000 years old – there is evidence for the use of mineral pigments for drawing and or colouring. There is also evidence of fishing, which suggests some novel technology or food-procuring techniques. The pattern may be one of gradual accumulation of 'modern behaviour', although many linguists doubt that syntactical language could develop gradually over time. Some anthropologists have suggested that a dramatic change – possibly driven by a genetic mutation related to speech – took place shortly before modern humans began to expand out of Africa.

Regardless of how and when modern humans acquired their language abilities and other elements of modern behaviour, they spread out of Africa no later than 50,000 years ago. Some modern humans are present in the Levant – on the doorstep of Africa – roughly 100,000 years ago, but their appearance seems to precede the main event. In fact, they may have been at least temporarily displaced by Neanderthals, who intruded into the Near East during the early cold phase of the last glacial period (about 60,000 years ago). However, modern humans show up in Australia 50,000 years ago. They were the first humans to cross the water barrier that separates the island continent from Southeast Asia – presumably by engineering some form of watercraft.

Analysis of the genetics of living human populations confirms that we are all either Africans or a subset of Africans that migrated into Eurasia in one or more waves. Between 50,000 and 45,000 years ago, modern humans returned to the Near East and expanded rapidly into southern Siberia and eastern Europe. Although their skeletal remains are extremely sparse in the earliest sites, their presence is revealed by characteristic types of artifacts and other traces of settlement that are unique to modern humans. On the central plain of eastern Europe, they camped at Kostenki – a group of open-air sites on the Don River. The oldest occupation levels are buried below a volcanic ash dating to about 40,000 years ago. These levels contain points and awls shaped from bone, digging tools made from antler, and a small carved piece of mammoth ivory that may represent an unfinished piece of sculpture; they also contain shells carried many hundreds of miles from the Black Sea area, indicating contacts over a much wider area than were achieved by Neanderthal groups.

The Kostenki sites reveal a pattern of adaptation to the local environment through imaginative innovation. Modern humans ventured into this region after the early cold phase of the last glacial when climates were oscillating between brief periods of mild warmth and cooler intervals. Levels below the volcanic ash contain large quantities of bone from small game such as hare, as well as some bird remains, and it appears that their occupants had created some new technology – snares, nets and throwing darts, perhaps – to harvest foods beyond the reach of the Neanderthals. Chemical analysis of a human bone above the volcanic ash indicates surprisingly high consumption of freshwater aquatic foods, suggesting more novel implements and devices.

The levels above the volcanic ash also contain the oldest known sewing needles in the world. Eyed needles of bone and ivory reflect the design and production of tailored clothing to cope with winter conditions that must have been a challenge to new arrivals from southern latitudes. There are traces of artificial shelters with interior hearths. Human skeletal remains – which are not sparse in these levels – underscore the need for this technology. They reveal an anatomy suited for the tropical zone and high susceptibility to cold injury in Ice Age Europe.

Equally impressive are the oldest traces of humans in the Arctic itself. Isolated sites in northernmost Eurasia reveal at least brief visits above the Arctic Circle between 40,000 and 30,000 years ago. A recently discovered site in northeast Asia near the mouth of the Yana River is far above the Arctic Circle at 71°N. The sites were presumably occupied during the summer months to exploit seasonally abundant resources, probably including migratory waterfowl. Their remote location is further evidence of the increased mobility of modern humans, who moved over great distances relative to their predecessors.

A recent recalibration of radiocarbon dates – which is necessary to adjust the dates to past fluctuations in atmospheric radiocarbon – indicates that modern humans spread rapidly across western Europe around 42,000–41,000 years ago. The Neanderthals were still present in this part of the world, and their encounter with modern humans has been the subject of much speculation and debate. The Neanderthals were making some bone implements and simple ornaments at the time, and many archaeologists believe that the pattern reflects some form of influence from incoming modern human groups. Others argue that the Neanderthals had begun to change before modern humans arrived. Possible interbreeding between the two populations has been inferred from Neanderthal traits reported on modern human skeletal remains from places like Oase in Romania. In any case, the Neanderthals were gone by about 30,000 years ago, leaving little mark on the people who inherited Europe and the rest of the world.

Modern humans moved into many of the caves that had been home to the Neanderthals. In the early years of the 20th century, archaeologists resurrected the culture – named 'Aurignacian' – of the first modern humans in western Europe from such caves in France, Spain and southern Germany. They found that the Aurignacian people had arrived during a period of intense cold – so cold that trees were scarce and bone was used as fuel in fireplaces. Tundra animals such as reindeer roamed the landscape in large numbers and provided much of their food. Despite the arctic conditions, the sewing needles found in sites further east are lacking in Aurignacian sites – needles show up later in western Europe when climates deteriorated further during the maximum of the last glacial – suggesting that the Aurignacians had no sewn clothing. And although the first modern humans in western Europe exhibit some innovations such as the split-base bone spear point (fitted securely to its socket with a clever wedging mechanism), their technology may have been uniquely primitive in comparison to later foraging peoples.

It is striking therefore that Aurignacian art was rendered with such sophistication and skill that it would not seem out of place in a 21st-century art exhibition. Only the subject matter – which includes animals that no human has seen for more than

above Tools and pigments such as these were used to create the spectacular cave paintings of southern France and northern Spain. The pigments, produced from ochre and charcoal, represented an early form of chemical technology.

below Although their age remains the subject of debate, the cave paintings at Chauvet in southern France are widely believed to date to the early Aurignacian period, more than 30,000 years ago.

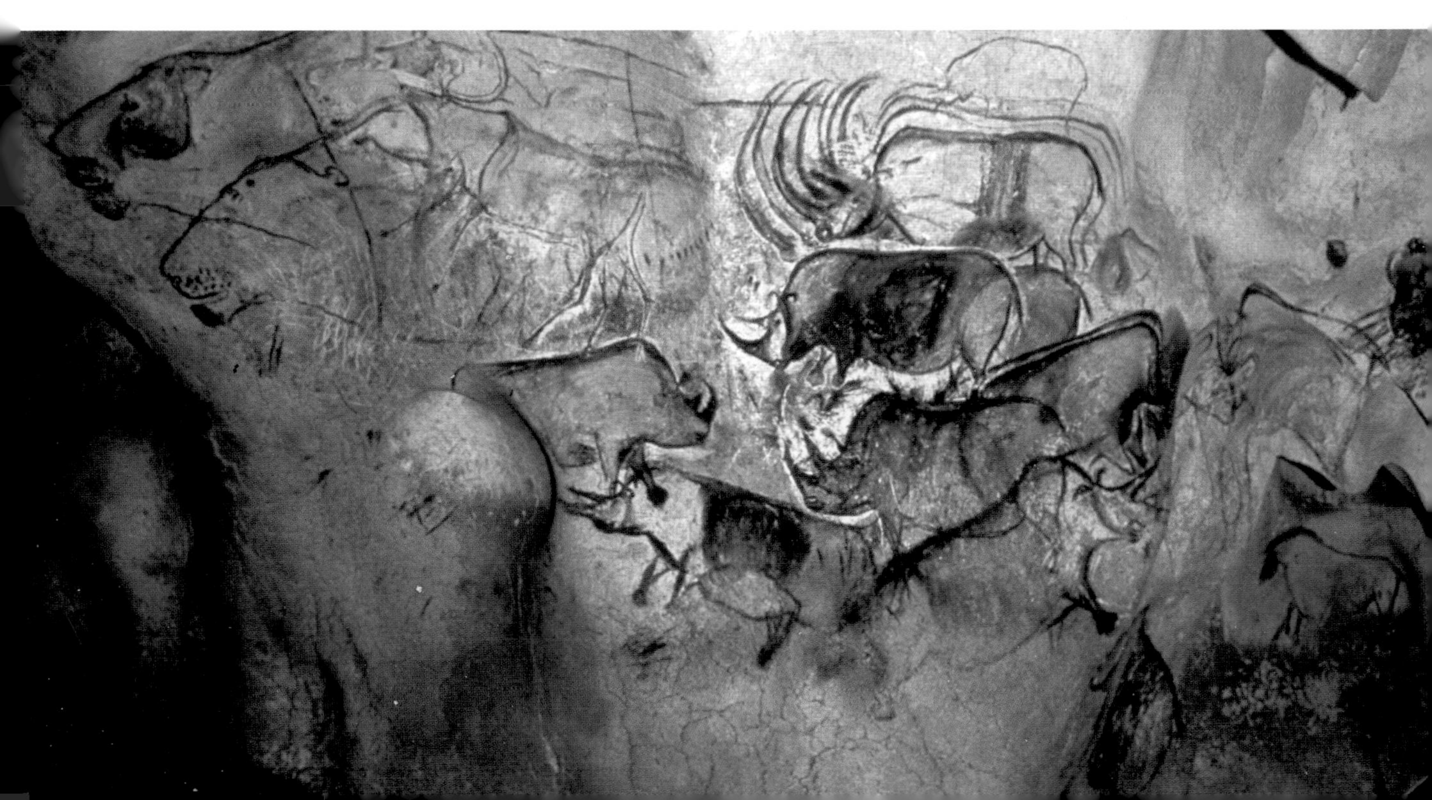

10,000 years – would appear strange. As recently as the early 1990s, some archaeologists remained convinced that Aurignacian art was as primitive as the technology. But the 1994 discovery and subsequent dating of paintings in Chauvet Cave in southern France ended debate on this issue. Using the accelerator mass spectrometry method of radiocarbon dating (which yields age estimates from minute samples of carbon), archaeologists directly dated a series of remarkable paintings of horses, woolly rhinoceros, and other subjects on the walls of Chauvet Cave to Aurignacian times.

Equally impressive are sculptures from Aurignacian sites in southern Germany. These include an elegant miniature horse and a curious human-like figure with

The complexity and sophistication of early Aurignacian art is illustrated by the Löwenmensch (meaning 'lion man') sculpture from Hohlenstein Stadel in Germany. This ivory figurine appears to represent an imaginary being – part human and part animal.

the head of a lion – both carved from mammoth ivory. The skill and imagination invested in these sculptures is comparable to that of the paintings, and in the absence of direct evidence for speech, the visual art of the Aurignacians provides critical evidence for language. The paintings and sculptures share fundamental properties with language: both manifest a potential for limitless creativity (each painting and sculpture is unique) and a complex hierarchical structure. Nor was Aurignacian creativity confined to the visual arts; their sites also yield the oldest known musical instruments in the world. An analysis of wind instruments (they may be classified as 'pipes') recovered from caves in France and Germany reveals a highly sophisticated design. But unlike the paintings and sculptures, we know nothing of the content of Aurignacian music.

Female figurines (often termed 'Venus figurines') are found in many European sites – especially those in central and eastern Europe – dating to the millennia preceding the last glacial maximum. This 25,000-year-old figurine is carved in chalk and was recovered from Kostenki in Russia.

The march of progress: the Upper Palaeolithic

The archaeological record for the millennia that followed the spread of modern humans across Eurasia and beyond provides evidence of a steady accumulation of ideas and knowledge, recorded in the material culture of the people who created it. This record is part of what archaeologists in the late 19th century termed the Upper Palaeolithic or Later Stone Age, although the terminology now seems anachronistic.

A wave of innovation and change ensued between 30,000 years ago, when the Neanderthals disappeared, and the maximum cold of the last glacial period about 23,000–21,000 years ago. During this period, human settlements in the cold landscapes of northern Eurasia reached unprecedented size and complexity. In southern Siberia, at sites like Mal'ta near Lake Baikal, people constructed large dwellings containing multiple fireplaces. In Russia, the remains of a semi-subterranean dwelling look similar to the winter houses of recent people of the north like the Inuit.

Other sites in Russia – including younger levels at Kostenki – reveal even larger settlements that must have been occupied by many families, at least for a period of some days or weeks. The settlements comprise a line of former hearths surrounded by pits of varying size. The numerous pits were probably used for storage of perishable materials, especially if these sites were visited during warmer months. Although the surface layer of the ground thawed during the summer, the soil below the surface remained frozen; digging down to the frozen layer, the settlers could create a natural refrigerator similar to the 'ice cellars' of the Inuit. Various foodstuffs – meat, fish, eggs and so forth – could be preserved, but another important function may have been keeping mammal bone fresh for fuel. Wood fuel seems to have been scarce in these landscapes and the former hearths are composed of burned bone and bone ash.

Roughly 25,000 years ago, large groups of people gathered periodically at places like Kostenki on the Don River in Russia, perhaps for reasons both economic and social. The deep pits surrounding the central line of hearths suggest that these gatherings took place during warmer months of the year when the earth was unfrozen.

During this period, people developed a number of new technologies and improved on existing technology, building on the innovations of the preceding interval, when modern humans were adapting to new environmental conditions as they moved into northern Eurasia. After 30,000 years ago, climates became colder as the Scandinavian ice sheet began to expand as it had during earlier glacial intervals. The approaching maximum cold of the last glacial period presented modern humans with a new set of environmental challenges as much of northern Eurasia became an arctic landscape.

Despite the cold and aridity, there was no shortage of food resources for the inhabitants of this landscape, especially given their talent for thinking up new ways to exploit those resources. The changing climate had brought arctic conditions to the middle latitudes where received solar energy remained high relative to the far north. The result was a bizarre mixture of tundra and grassland that was significantly richer in plants and animals than the tundra of the Arctic today. Reindeer and musk ox lived alongside bison and horse and sometimes elk and red deer. Birds, including migratory waterfowl, seem to have been common in some areas and freshwater fish were apparently available.

It is not known if population density (i.e. number of persons per square kilometre) increased at this time, but it is clear that the size and complexity of the settlements had expanded. The pattern suggests at least temporary gatherings of unprecedented scope, perhaps reflecting temporary aggregations of families at specific times of the year when particular food resources – such as fish or reindeer – were concentrated in one place for a brief interval of time. Such aggregations must have had social as well as economic significance – an opportunity to reinforce ties among families through feast, festival and marriage. It is a pattern found among many recent foraging peoples.

By 30,000 years ago, people living in parts of northern Eurasia had engineered the technology of fired ceramics and began to produce fired-clay objects in kilns. As far as we

below An anthropomorphic figurine of fired clay from the Maininskaya site, which is located on the Yenisei River in southern Siberia, is dated to about 18,000 years ago.

can tell, the objects – which included female figurines – had no 'practical' value: they were created to express ideas about the world and perhaps for ritual performance. Other advances in pyrotechnology include portable lamps, fuelled with animal fat and equipped with lichen wicks, and a possible baking oven from the site of Ohalo II in Israel, which is dated to about 22,000 years ago.

At the same time, cave paintings of western Europe reveal advances in chemical technology. The paints used for these two-dimensional images were synthesized from organic and inorganic components. Information technology is represented by bone fragments engraved with rows of markings that record data about some aspect of the environment. Some of these have been interpreted as lunar calendars – the first evidence for the structuring of time – similar to those made by recent foraging peoples. And the first mechanical technology – instruments and devices composed of moving parts – is present by the later phases of the last glacial maximum, if not before. The earliest known example is a spearthrower from a site in France dating to almost 20,000 years ago.

Despite these impressive achievements, conditions at the height of the last glacial – roughly 23,000–21,000 years ago – may have been too much for people in the coldest and driest parts of northern Eurasia. There seems to have been a hiatus of settlement in these areas, lasting perhaps a thousand years or more until climates began to ameliorate after 20,000 years ago. At the site of Molodova, for example, in southwestern Ukraine, the continuous sequence of occupations – otherwise spanning the entire Upper Palaeolithic – is interrupted for several millennia. Further east at Kostenki, settlement also terminated at this point, and the number of sites in Siberia declined significantly during this interval.

Why couldn't people cope with the conditions of maximum cold during the last glacial? The answer is not obvious. Modern humans had engineered an array of ingenious technologies in the Ice Age landscapes of northern Eurasia, and they seemed to have thrived in these environments until this point. Perhaps the density of available resources – food and/or fuel – fell below a critical threshold after 23,000 years ago in some areas. Perhaps the people's anatomy (which still retained the proportions of peoples now living in the tropical zone) left them vulnerable when winter temperatures declined dramatically. It may be significant that skeletal remains of the people who subsequently reoccupied these areas exhibit a reduction in limb dimensions.

The earliest known examples of mechanical technology (i.e. implements or devices with moving parts) date to the later Upper Palaeolithic and are represented by spearthrowers, such as this example from Bruniquel in southwestern France.

Humans at the end of the Ice Age

The last 5,000 years of the Ice Age was a time of apocalyptic climate change, on a scale far greater than anything so far witnessed today, and major human migration. It was also a period of continuing innovation and culture change that seems to have anticipated the trend towards permanent settlements, agriculture and civilization that took place after the Ice Age. During the final millennia of the Ice Age, we see the first signs of pottery, animal domestication and what appear to be semi-permanent camps. Most importantly for human geography, people expanded across the so-called Bering land bridge, shortly before rising sea levels flooded it, and into the New World.

One of the most striking effects of the growth of the massive glaciers in the Northern Hemisphere was the corresponding fall in sea level. When the glaciers increased in size, they locked up billions of gallons of water in the form of ice and the oceans shrank, exposing continental shelf areas around the world, as we saw in Chapter 4. The shallow sea bed between northeast Asia and Alaska – now the Bering Strait – became a dry plain connecting the continents of Asia and North America. Once people learned how to live in high-latitude environments, they could cross the land bridge into the Western Hemisphere.

Across Europe, the final stage of Ice Age culture is known as the 'Magdalenian' epoch. It is associated with spectacular cave paintings at Lascaux and other classic sites in France and Spain, but it is the technology of this era that provides a more significant contrast with the earlier cultures of the Upper Palaeolithic. The pace of innovation and change – the march of progress – continued and perhaps accelerated. Evidence of the first bows and arrows, an ingenious piece of mechanical technology, appear no later than 14,000 years ago. The evidence includes wooden arrow shafts discovered many years ago in Germany; arrowheads are reported from an even earlier site in Japan. By storing and controlling the release of energy with a drawn bow and string, hunters could launch a projectile with greater power and accuracy than if it was simply thrown. Use of the simpler spearthrower continued, but there seem to have been some improvements in design. Barbed harpoons and other implements for harvesting fish were widely used as well.

below left Barbed harpoons appear in the later Upper Palaeolithic of northern Eurasia and reflect an increasing emphasis on aquatic resources.

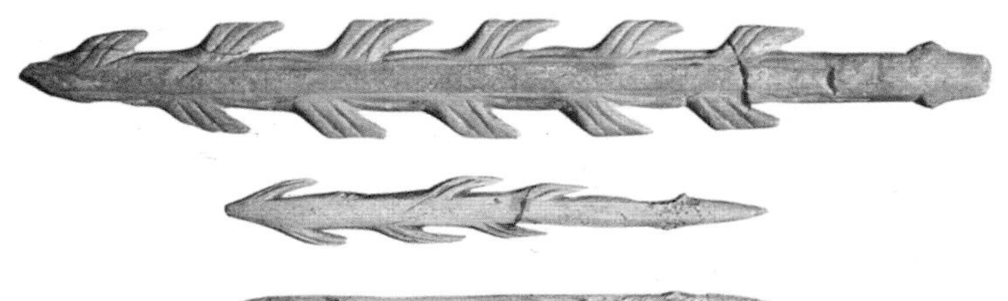

Perhaps the increasing efficiency of the food-procuring technology allowed for larger groups and reduced the need for mobility. During the final millennia of the Ice Age, more permanent-looking settlements appear; they seem to anticipate the villages of the postglacial epoch and the sedentary way of life that ultimately led to urban centres and civilization. At Plateau Parrain in the Dordogne region of France, traces of former houses with paved stone floors are found grouped together in a village-like setting. The economy retained an Ice Age flavour, however, with continued hunting of reindeer and mammoth. On the central plain of eastern Europe, the settlements also reveal a village-like character, with individual houses assembled from mammoth bones and tusks. At Mezhirich in Ukraine, four of these

below Dwellings constructed with mammoth bone and tusk, such as the one reconstructed below, are known from a number of sites in eastern Europe (where wood was probably scarce), occupied during the millennia following the last glacial maximum.

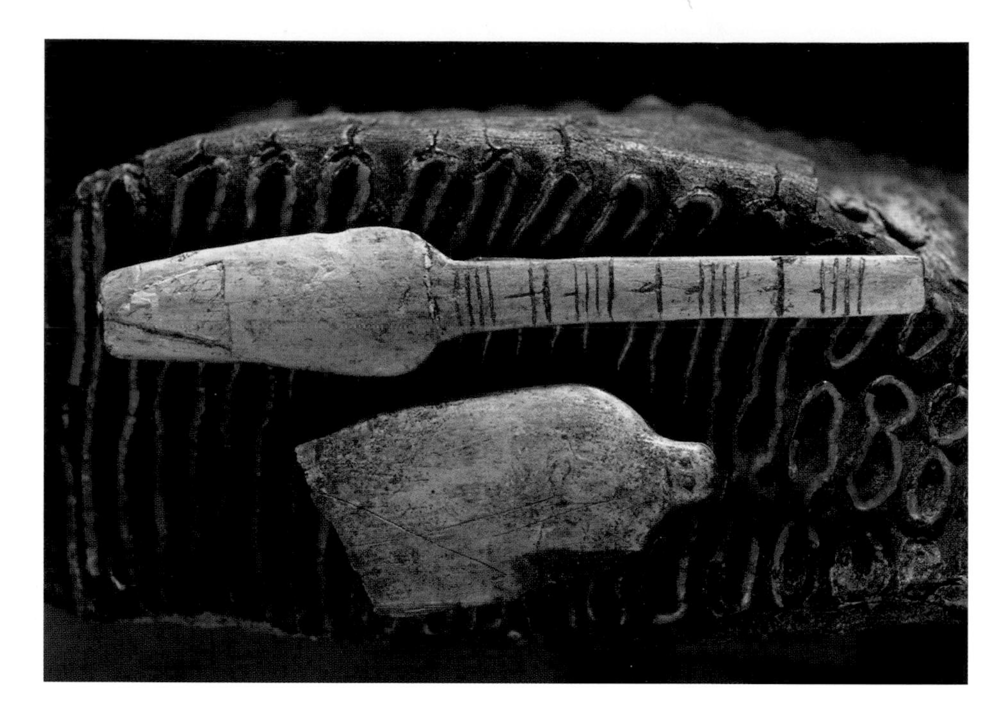

previous pages The famous cave paintings at Lascaux in France are dated to the later Upper Palaeolithic. This image appears to depict a bison, rhinoceros, and hunter lying next to a broken spear – the only human figure known anywhere at Lascaux.

left Mezhirich, which is located in Ukraine and dates to about 18,000 years ago, contained four mammoth-bone house ruins and many examples of carved and engraved art objects.

remarkable structures, dating to roughly 18,000 years ago, were found in a group – the oldest known 'ruins' on earth. The use of mammoth bones for construction presumably reflects a continuing shortage of wood on the central east European plain – still a cold dry steppe – and fire pits inside the houses were filled with bone ash.

Another sign of declining mobility may be the appearance of pottery vessels. Pots are heavy, and their presence suggests less movement from one settlement to another. The earliest known ceramic pots date from the final phase of the Ice Age. Although fired ceramic technology had been developed much earlier, its use for production of water-tight containers was a novelty. Pottery vessels are found in the Far East – both in Japan and on the Asian mainland – as early as 16,000 years ago.

Also suggestive of more permanent settlement is the appearance of domesticated dogs. At sites in Russia and Germany, the remains of dogs are dated to 18,000 and 16,000 years ago. A study of the mitochondrial DNA of modern dogs indicates an origin in East Asia as early as 20,000 years ago. The combination of dogs and bow-and-arrow technology must have brought human hunting to a new level of efficiency. In the Near East, settlements comprising semi-subterranean houses of stone appear by 15,000-14,000 years ago and mark the beginning of the shift towards village agriculture. For people in this part of the world, the Ice Age was already over.

As the immense ice sheets in northwest Europe and northern North America retreated, people began to expand into places that had been abandoned during the coldest phase of the last glacial period. Groups spread into recently deglaciated areas

By late Upper Palaeolithic times, if not earlier, people in northern Eurasia had domesticated dogs, and biotechnology may be added to their impressive array of innovations.

and places at the margins of the ice in northern Europe after 15,000 years ago. In other parts of northern Eurasia, people moved into regions that, while not glaciated at the height of the last glacial period, had, nevertheless, been largely or wholly uninhabitable. The most important of these was probably northern Siberia.

The large sites with houses like Plateau Parrain and Mezhirich were absent in Siberia. The interior of northern Asia remained a dry and cold environment where the availability of sources of food for humans in any given area was low. People were evidently scattered across the region in small highly mobile groups that relied primarily on the hunting of large mammals. They gradually expanded northwards – probably following the return of trees into higher latitudes after 16,000 years ago. A small occupation at Dyuktai Cave on the Aldan River in eastern Siberia is dated to about 15,000 years ago.

From places like Dyuktai Cave, the Siberians moved north and east into the western half of a vast land that stretched across the Bering Strait region to the Yukon. One day roughly 15,000 years ago, someone scrambled up an embankment – now the western coast of Alaska – and stepped unknowingly into the Western Hemisphere. It was a major achievement for humankind, which until that moment had been confined, along with its African ape ancestors, to the other half of the earth. Only modern humans with their ability to remake themselves and their surroundings had been able to overcome the environmental barriers to the New World. At the same time, it was the effects of Ice Age climate on sea level that had created the land bridge to the other hemisphere.

By 14,000 years ago, people had settled in the Tanana Valley of central Alaska. The cold steppe that had supported herds of mammoth and horse was being rapidly transformed into a shrub tundra environment. Grasses were replaced with shrubs that offered little sustenance for large grazing mammals, many of which began to decline in numbers. Animal remains from one of the Tanana Valley sites provide a glimpse of a way of life that had already acquired a post-Ice Age quality. Small

By 14,000 years ago, people were camping on the Tanana River in central Alaska and harvesting a rich bounty of waterbirds, in addition to hunting large mammals. Their hearths were largely fuelled with bone.

mammals and birds, especially waterfowl like tundra swan and mallard, predominate. Aquatic habitats had proliferated in a rapidly thawing landscape. The large mammals are chiefly represented by postglacial species like elk, but some isolated bones of horse and mammoth are also present. The artifacts are virtually identical with those made at Dyuktai Cave and tie the early Alaskans to the interior of northeast Asia. Wood remained scarce and people continued to burn bone in their hearths.

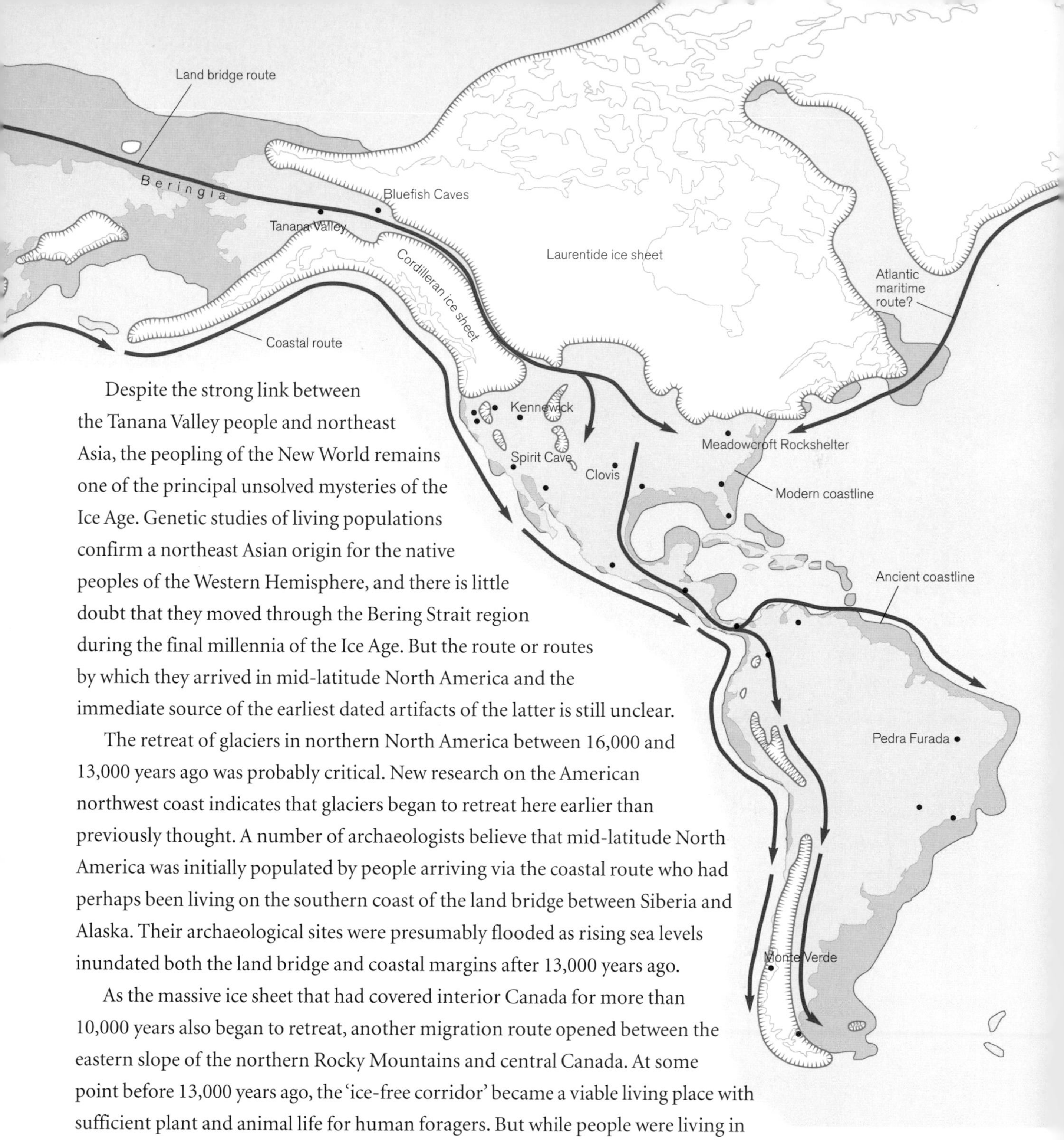

Despite the strong link between the Tanana Valley people and northeast Asia, the peopling of the New World remains one of the principal unsolved mysteries of the Ice Age. Genetic studies of living populations confirm a northeast Asian origin for the native peoples of the Western Hemisphere, and there is little doubt that they moved through the Bering Strait region during the final millennia of the Ice Age. But the route or routes by which they arrived in mid-latitude North America and the immediate source of the earliest dated artifacts of the latter is still unclear.

The retreat of glaciers in northern North America between 16,000 and 13,000 years ago was probably critical. New research on the American northwest coast indicates that glaciers began to retreat here earlier than previously thought. A number of archaeologists believe that mid-latitude North America was initially populated by people arriving via the coastal route who had perhaps been living on the southern coast of the land bridge between Siberia and Alaska. Their archaeological sites were presumably flooded as rising sea levels inundated both the land bridge and coastal margins after 13,000 years ago.

As the massive ice sheet that had covered interior Canada for more than 10,000 years also began to retreat, another migration route opened between the eastern slope of the northern Rocky Mountains and central Canada. At some point before 13,000 years ago, the 'ice-free corridor' became a viable living place with sufficient plant and animal life for human foragers. But while people were living in this region 12,000 years ago, their presence at an earlier time has yet to be confirmed.

Ironically, the oldest widely accepted evidence for New World settlement is found in southern South America at the site of Monte Verde in Chile. The artifacts are few and rather difficult to relate to the archaeological record in northeast Asia and Alaska. Somewhat later sites are widespread in North and South America,

The initial peopling of the New World seems to have occurred no later than 15,000 years ago and may have entailed migration along the Pacific coast of North America.

The distinctive fluted Clovis spear point is found widely in North America and dates to before the Younger Dryas cold event. It was used for the killing of large mammals, probably including mammoth.

including the well-known Clovis sites of North America, which date to 13,000 years ago and slightly before; these sites contain stone points with sides fluted to facilitate hafting onto spear shafts, and sometimes the remains of mammoths – by now rapidly approaching extinction in most parts of the world (see Chapter 6) – killed or scavenged by their occupants. Steppe bison soon became the primary large-mammal focus of people living on the High Plains as the Ice Age came to a close.

The final gasp of the Ice Age occurred in the form of a severe cold snap that took place about 12,000 years ago. This so-called Younger Dryas event brought cold and dry conditions back to many parts of the Northern Hemisphere as temperatures dropped by several degrees. Perhaps its most bizarre consequence was the appearance of bison-hunting people from the High Plains in Yukon and Alaska – apparently moving up through the ice-free corridor to exploit expanded bison populations in the far north. The phenomenon was short-lived. Within a few hundred years, climates warmed again and the bison-hunter camps, with their characteristic artifacts of the High Plains, vanished from the Arctic.

The Younger Dryas event is also associated with the shift towards settled life and agriculture in the Near East. Archaeologists speculate that the brief cold snap forced some changes in the subsistence economy that underlie the rapid emergence of village farming. Within a few millennia, the Near East saw expansion of agriculture, growth of urban centres, and civilization. The process was repeated in other parts of the world. Earlier humans – including the Neanderthals – had confronted climate changes similar to those at the end of the Ice Age, but without a comparable response. The difference lies in the creative powers of the mind of modern humans and their ability to reshape themselves and their environment.

6
The Ice Age Bestiary

IN THE COOL LIGHT OF A GLACIAL MORNING over 130,000 years ago, a group of Neanderthals are butchering mammoths at the foot of a cliff. These are the prized remains of a small herd of females and their young that have been stampeded, terrified and trumpeting, over the cliff edge to provide food for the hunting band and their families. The place will become La Cotte de St Brelade, a cave site on the island of Jersey, but for now it is a promontory on a wide plain that stretches from France to England and is dotted with small groups of rhinos, horses and humans.

This was a very different world from the one in which we live now, and the animals of today are the impoverished remnants of a huge diversity of creatures, such as the sabre-toothed cat, woolly rhino and marsupial lion, that existed until approximately 11,000 years ago. They were the product of millions of years of evolution but failed to survive the great extinction that occurred at the end of the last glacial. This was not the first major extinction that the world had seen – the loss of the dinosaurs 65 million years ago is perhaps the best-known – but it is the most recent, and it may also mark the first occasion that modern humans really began to have an impact on our planet.

previous pages The Columbian mammoth, *Mammuthus columbi*, was the largest of all the mammoths. They inhabited North America and Mexico in the later Ice Age.

The distribution of animals

The earth is constantly changing and even features as large as seas or as vast as the continents have not always been where we see them today (Chapter 3). Such changes have affected the distribution of animals, although it is not always physical barriers that dictate where an animal can live. Some, such as the reindeer (caribou) and musk ox are restricted by temperature or vegetation to the northern steppe, which stretched, during glacials, from northern Spain, through Russia and Alaska to the west coast of North America. Others like kangaroos and koalas have evolved in isolation following the movement of the earth's crust through plate tectonics, which made Australia an island continent over 35 million years ago. Movements of the earth's crust also caused the Mediterranean Sea to dry up by 5.5 million years ago, and it reflooded some 200,000 years later. During this time it would have been possible for animals to walk from North Africa to Europe across the vast salty desert that is now the modern sea bed. Several animals that started this walk became trapped on islands when the seas rose, evolving into species found nowhere else, such as the strange goat-like animal called *Myotragus* from Majorca. Other animals found on islands, like dwarf elephants and flightless swans, reached them under their own steam, by flying, swimming or rafting, and evolved *in situ*.

In another example of continental movements, around 3 million years ago North and South America joined together enough at the Isthmus of Panama to allow the animals that had evolved on each continent to disperse into the neighbouring lands.

left The musk ox, *Ovibos moschatus,* was a familiar sight in glacial periods. Now confined to North America, in the past they were found in northern Eurasia during glacial periods. They have very dense fur coats to protect them from the cold, and when threatened by predators the adults form a defensive line or, if the group is large enough, they form a protective ring around the juvenile members of the herd.

below Remains of the extinct *Myotragus* were first found and published by Dorothea Bate in 1909. *Myotragus* evolved on Majorca and Menorca, and unique among the sheep and goat family, it had two large front incisors that were perpetually growing, like those of a rodent.

This has been termed the Great American Interchange and at this time many animals, including ground sloths, anteaters and armadillos, moved north, and sabre-toothed cats, tapirs and horses moved south. To put this dispersal into historical perspective, by the time of the Great American Interchange human ancestors had been walking around Africa for at least 3 million years, but it was to be another million years before they began intercontinental movements of their own.

Giant Mice and Dwarf Elephants

Islands around the world have produced giants and dwarves of a variety of species, ranging from surprisingly large mice to tiny elephants. Changing sea levels can isolate animals on islands, or they can swim or raft there. There is an evolutionary trend on islands for small animals to get bigger as there are usually few predators, and for large animals to get smaller as there is a limited amount of vegetation to support them. Some animal families, such as the elephants, have dwarfed repeatedly: dwarf mammoths have been found on Wrangel Island in northern Russia, St Paul in the northern Pacific and the California Channel Islands, while dwarf stegodons are reported from islands in Southeast Asia. Dwarf straight-tusked elephants have been found in the Mediterranean, and those from Crete tell a tale of repeated colonization. Crete in the early Ice Age had pygmy elephants and hippos and a giant mouse the size of a brown rat. Later, mainland-sized elephants colonized the island, along with deer which evolved rapidly into dog- and moose-sized forms with at least four others in between. The Mediterranean dwarf elephants were smaller than a family dining table, and the juveniles were the size of cats. The majority of animals that are found on the Mediterranean islands are those that are capable of swimming and move in herds, such as elephants and deer, while sloths managed to colonize many of the islands in the Caribbean. Several animals arriving at once makes the chances of successful colonization much more likely. Small islands naturally have limited space available, so these populations were never abundant and the arrival of a new species such as brown rats or humans has led to massive extinctions of these unique faunas all over the world.

right Dwarfed forms of straight-tusked elephants, *Palaeoloxodon antiquus*, have been found on several of the Mediterranean islands. Why they were so small is not well understood, but possible reasons are shortage of food and a lack of predators on islands.

opposite A skeleton of an extinct species of pygmy hippopotamus from Madagascar. Dwarfed forms of hippopotamus have been found on several islands, as they are capable of swimming for some distance in salt water.

below Map of the Mediterranean showing the dwarf and giant species found on the islands.

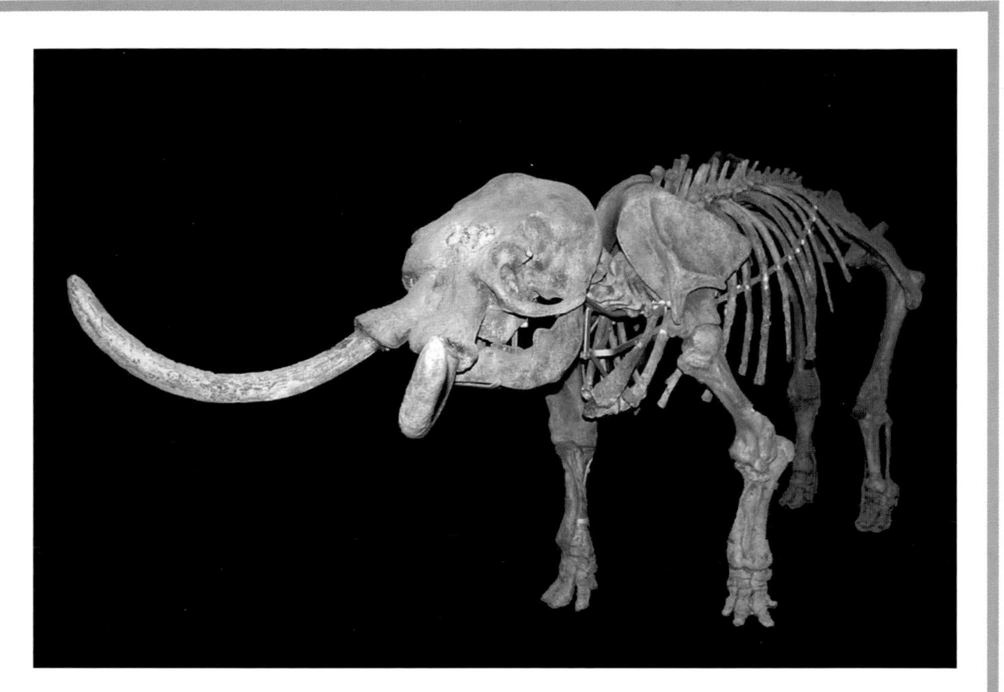

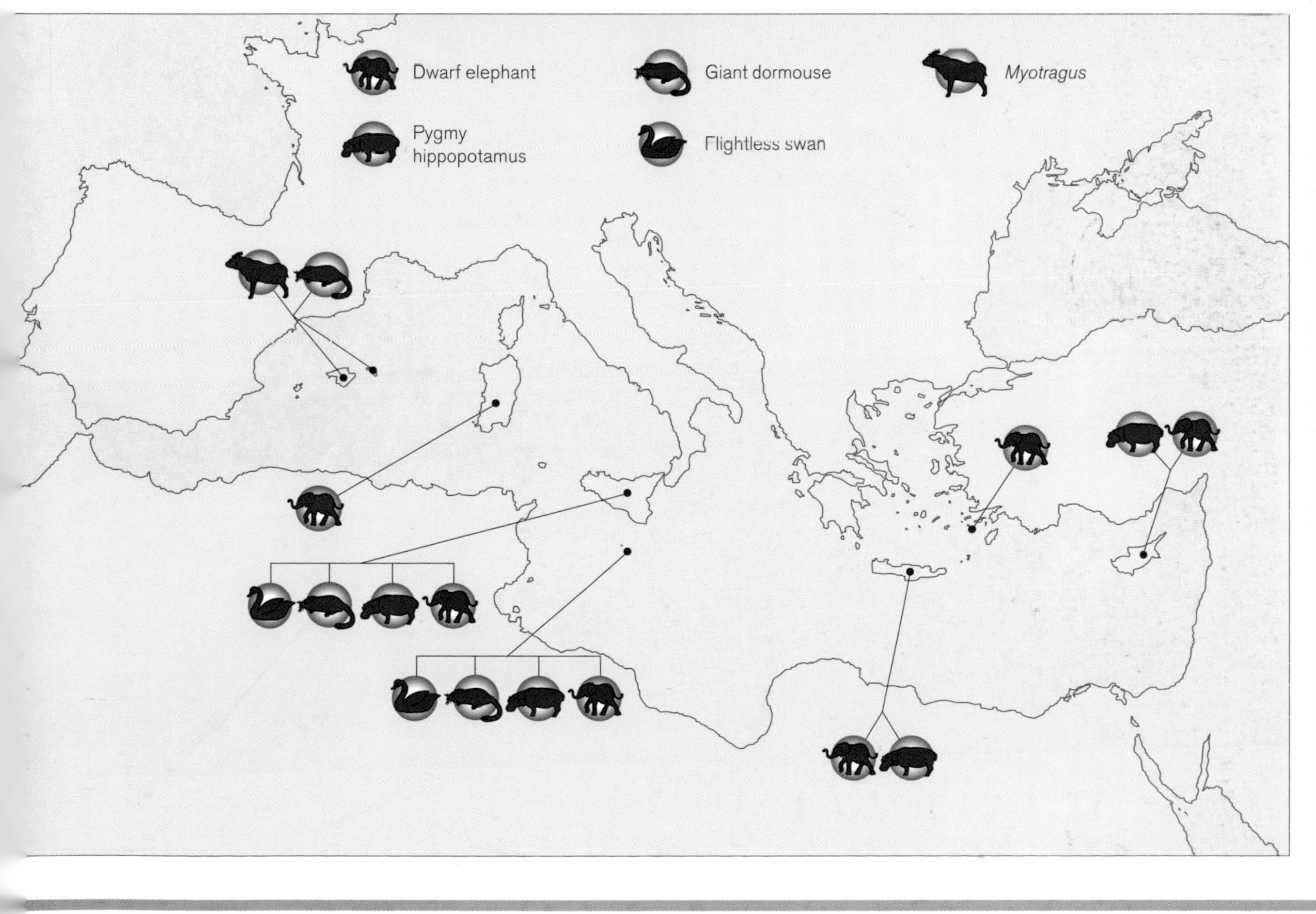

Inside and out of Africa

Hominins first left Africa some 1.8 million years ago, and there have been several dispersals since. Movement of animals from one continent to another happened repeatedly, as new forms evolved and land bridges created by tectonic movements or low sea levels in glacial periods gave them the opportunity to disperse into new areas. The woolly mammoth, an animal that is almost synonymous with the Ice Age, had an African origin, as the first members of its family have been found in deposits in southern and eastern Africa dating to 5 million years ago. It appears that later members of the mammoth family dispersed out of Africa over 3 million years ago and began to evolve into the steppe-dwelling animal that we would recognize, not in Africa, but in northern China and Siberia. In contrast, some animals that are considered African today in fact had their origins in the Americas, including the zebra, whose ancestors dispersed across the Bering land bridge into the Old World over 3 million years ago, and also possibly the cheetah.

As the climate cooled at the start of the Ice Age, a major change occurred in the African fauna. The climatic changes had a direct impact on the plants, through a reduction in moisture and temperature, which led to the spreading of grasslands and an overall decrease in forested areas. Once the environment began to change from forested to more open areas, antelopes, pigs and primates that were adapted to eating fruits and leaves would have been at a disadvantage. These foods are relatively soft, while grasses contain a lot of silica, which is gritty and causes teeth to be worn down more quickly, resulting in animals not being able to feed effectively so they often starved to death; grass is also harder to digest. Over time some species of pig developed larger teeth with thicker enamel that were more resistant to wear, while antelopes such as the wildebeest developed longer tooth crowns that took longer to wear down to the gums. The newly arrived one-toed horses were already grazers and took advantage of the grasses straight away, quickly spreading across the continent to northern and southern Africa. Other browsing animals dispersed out of Africa, such as the ancestors of the straight-tusked elephant that moved into the Near East and Europe around 1 million years ago.

For some monkeys and browsing antelopes these changes were too much, as the loss of trees that they depended on for food and shelter and the changing adaptations of their competitors resulted in their extinction. And it was not just the herbivores that suffered, as solitary predators such as sabre-toothed cats, which probably used trees to hide behind before ambushing their prey, also became extinct. Predation, and particularly scavenging, in the open plains is much harder than in thickets where food is not so readily seen by competitors such as vultures, hyaenas and jackals. Many carnivores on the plains live in groups to help them hunt and defend their kills from each other. As hominins evolved they would have begun to compete with these large carnivores for food, but for much of our evolution we were most definitely prey, as finds from South Africa have shown.

A small group of robust australopithecines, a male and his harem of females, were foraging for food in the savannah. Suddenly there was a commotion in a small thicket where a young male had been collecting fruits. A sharp scream and a flash of a spotted coat, and it was all over. A leopard had been watching the group and waiting for its chance. Having pounced and killed the hapless hominin, the leopard firmly anchored its teeth in the eye sockets and scalp of its victim and dragged it

Palaeontologist C.K. (Bob) Brain and the skull cap of an unfortunate australopithecine from Swartkrans Cave, South Africa. These skull fragments still bear the holes caused by a predator's teeth when it grasped its prey. A leopard's jaw from the same site fits the toothmarks perfectly, providing evidence that hominins were prey rather than predator.

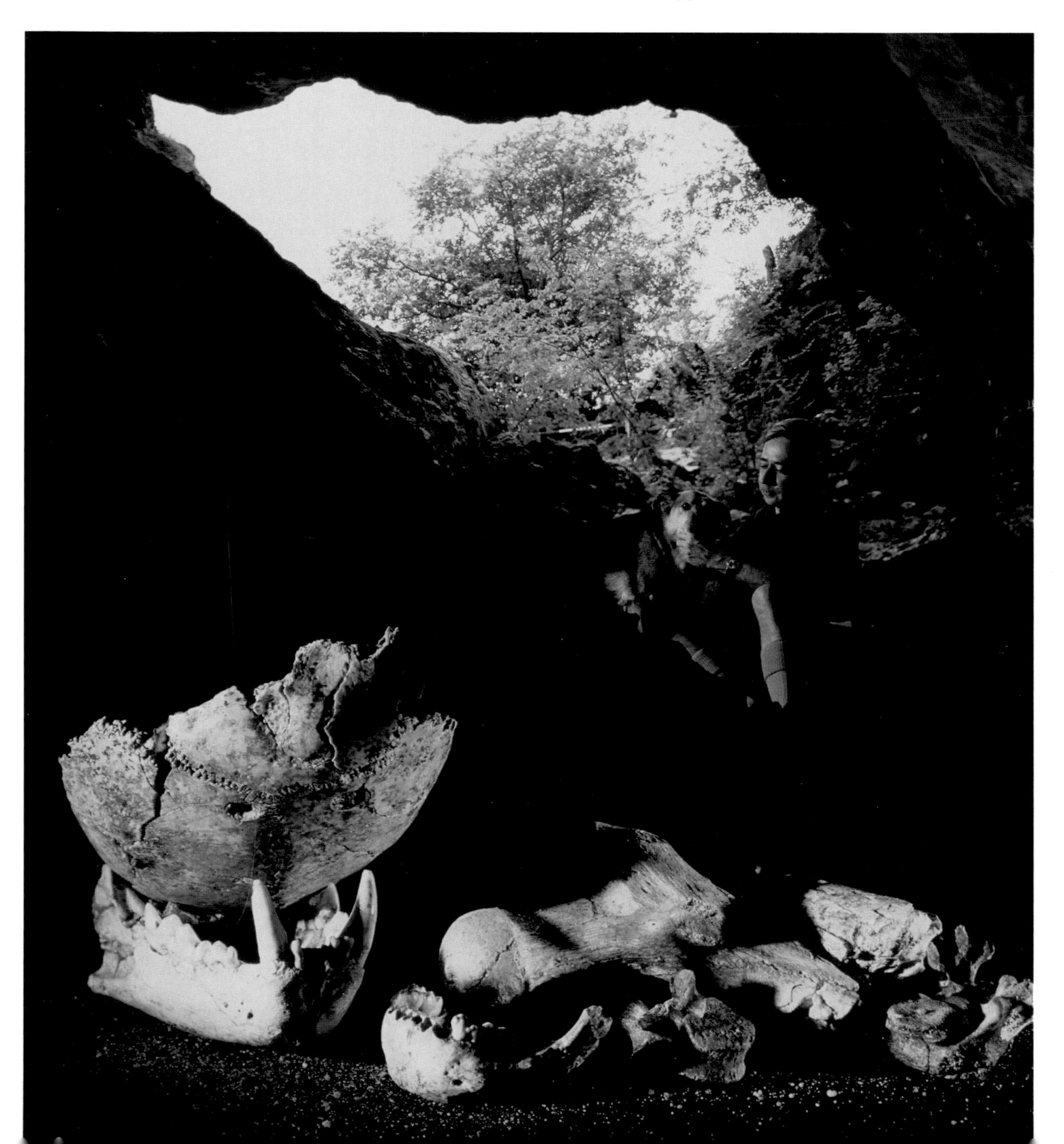

away to eat. This leopard chose to cache its food in a tree above a cave mouth to avoid scavengers, and from here a skull fragment fell into the cave, where it was excavated by palaeontologists around 1.5 million years later. The tooth marks left on the skull fitted the canines of a fossil leopard that was also found in the cave, eloquently revealing the story of life and death in the savannah.

The robust australopithecines and many other prey animals and carnivores were extinct by 1 million years ago, and there has been no major African extinction event since. In fact Africa is now the home of approximately one quarter of all living mammals. Although some animals such as the long-horned buffalo and the giant cape zebra disappeared at the end of the last glacial, the majority of large species survived, unlike those of all of the other inhabited continents. Such relative stability

There were many different species of rhino in the past, but only five remain today, two in Africa and three in Asia. They are all herbivores, but can be divided into grazers (grass eaters) or browsers (leaf eaters). The same is true for the extinct Ice Age rhinoceroses, as the woolly rhino was a grazer, while Merck's rhino was a browser.

in comparison with the other continents has led some people to suggest that the modern African ecosystem contains the only Ice Age megafauna left on the earth. It has also been suggested that there were few extinctions in Africa because the animals there had evolved with humans, and were not so naïve when faced by a human hunting party as those animals that had never seen one before. However, Africa was not the only place in which animals co-existed with humans for hundreds of thousands of years, as Europe had also been occupied by different species of human for over 1 million years before the present.

Europe – land of the hairy

If we were to watch a film of the last half million years of European history, what would we see? In any particular year, perhaps little change, maybe a glacier on a mountaintop waxing or waning as the snowfall begins or melts. But put the same film on fast-forward, and massive changes would occur – huge sheets of ice spreading from the Arctic across the ocean towards northern Europe; the North Sea and English Channel drying up; the cold grey tundra and steppe vegetation expanding in front of the ice sheets and the green temperate forests retreating south. Then, in a few seconds, things would reverse, the ice would retreat, the forests expand and the English Channel would once again separate Britain from the continent. These massive changes happened five times in the last 500,000 years and would have had an impact on all the plants and animals that were living in Europe at the time. But what were those animals? And what impact did the climate have?

The animals that inhabited Europe over the last half a million years can be divided into two main groups: the warm-adapted or temperate species (those that lived all over Europe in the interglacials, and survived south of the Alps and Pyrenees in the glacials) and the cold-adapted species (those that inhabited the steppe and tundra and moved into Europe in the glacial periods).

As the climate cooled into a glacial the vegetation changed from forested to more open environments, bringing with it a change in fauna, from those animals adapted to eating trees and shrubs, to those that graze. First the numbers of wild cattle, bison and horses would increase, and then, as the climate got colder, the mammoth steppe grasslands of low grasses, sedges and dwarfed trees would spread from Siberia out across Eurasia as far as the Pyrenees. With it would come the animals that had evolved upon it, the reindeer, saiga, woolly mammoths and woolly rhinos, as well as smaller animals such as lemmings, Arctic foxes and the wolverine. While the cooling into a glacial may have taken many thousands of years, the change to interglacials may have been very abrupt, and the cold-adapted mammals would die or retreat back to the east until the expansion of the steppe in the next cold phase. In their

The woolly rhinoceros, *Coelodonta antiquitatis*, was a cold-adapted creature, found in the northern parts of Europe and Asia. The head and neck of the woolly rhino were held fairly low, as shown in the reconstruction, indicating an animal that was adapted to grazing on low-growing plants.

place, first the trees and then the warm-adapted animals such as the narrow-nosed rhinoceros, Barbary macaque and hippopotamus would spread north from their refuges in Spain, Italy and the Balkans and cover the continent. However, not all animals present in the north went into refuges during the glacials – some stayed and these included cave bears, lions, spotted hyaenas and humans. The last three needed meat to survive, and as long as it was present it did not matter to them whether it was a warm- or cold-phase animal that they were eating.

Heidelberg people, the first humans to occupy northern Europe, lived alongside a wide array of carnivores. They would have known and feared the sabre-toothed cats and giant short-faced hyaenas that stalked the plains of Europe. These hyaenas were enormous, over 1 m (3.3 ft) high at the shoulder, with jaws up to 24 cm (9.5 inches) long – the length of an adult's arm from elbow to wrist. Like the modern spotted hyaena, these jaws contained teeth that were specialized for crushing the bones of their prey to extract the maximum nutrients from the carcass. There was also a large cheetah, some 50 per cent larger than the modern cat, but this was rare and a Heidelberger may never have seen one in their lifetime. By the time the Neanderthals had evolved in Europe the sabre-toothed cat, cheetah and short-faced hyaena had gone, and only the lion, leopard and spotted hyaenas were left.

Although the sabre-toothed cat was thought to be extinct in Europe by 300,000 years ago, a jaw of one of these cats was recently found in the North Sea and dated to approximately 28,000 years ago. These cats still lived in North America, so is this

Woolly rhino	
Genus	*Coelodonta*
Species	*Coelodonta antiquitatis*
Height	1.7 m (5.6 ft)
Period	200,000–10,000 BP
Location	Northern Eurasia

This woolly rhino skull shows the beast's immense frontal horn, probably used for duelling with other males, as well as the flattened molar teeth with which it grazed on grass. The woolly rhinoceros roamed about alone or in small family groups like present-day African rhinoceroses.

find evidence that they dispersed back into Europe in the last glacial, or did they survive in some currently unknown area in Europe? We do not yet know, but the find illustrates that palaeontology is a science of discovery and surprises.

When modern humans arrived in Europe, the large carnivores that were present would have been familiar to them, as they had evolved alongside their close relatives in Africa. But one species would have been new – the cave bear. There are no bears in modern Africa, although there were some millions of years ago, and the brown bear only died out in north Africa in historical times. The cave bear evolved in Europe and central Asia, ultimately becoming a very large vegetarian beast relying on hibernation to see it through the cold northern winters. Thus it would not have preyed upon humans, but it may have been a formidable opponent if they were in competition for caves or food resources. Some caves have thousands of cave bear bones within them, showing where the bears lived and died, often during hibernation (shown through wear on teeth and the estimated age at death of the animals).

The Eurasian cold-phase animals are the iconic Ice Age beasts such as the woolly mammoth, with a thick coat of hair and tiny ears and tail, and the woolly rhinoceros with a long hairy coat to keep out the cold and two horns, one of which could grow to over 1 m (3 ft) in length. Yet there were also temperate-adapted members of the elephant and rhino families. These were the straight-tusked elephant and Merck's rhinoceros – animals adapted to eating leaves off trees and bushes rather than the grasses and low shrubs that were consumed by the woolly rhino and mammoth.

Cave Bear	
Genus	*Ursus*
Species	*Ursus spelaeus*
Height	1.2 m (4 ft)
Period	300,000–10,000 BP
Location	Europe and western Asia

above The cave bear, *Ursus spelaeus*, had a very distinctive profile. The skull was domed above the eyes, with a steep drop down to the muzzle.

opposite Cave bear remains have been found in vast quantities – bones from an estimated 30,000–50,000 individuals were found in just one site. Here cave bear skulls are shown as they were found (inset) on the floor of Chauvet Cave in France (main picture), a site famous for its magnificent paintings of Ice Age animals.

Why was there the need for all these massive changes in fauna? Why did the mammoths and rhinos not just stay where they were while the climate changed around them, as humans and lions did? The answer is that elephants, rhinos and the rest are herbivores and they have to follow the vegetation to which they are adapted. Rather than a mass movement of animals, slow population expansion into new areas as suitable habitat spread is likely to have occurred. As the habitat contracted again, some animals would have followed it but many more would have become isolated in small populations in pockets of vegetation and would have eventually died out.

The straight-tusked elephant was a huge forest-dwelling animal standing up to 4 m (13 ft) high at the shoulder, which probably would have browsed trees by

The Columbian mammoth, *Mammuthus columbi*, and the American mastodon, *Mammut americanum*, had very different body plans, as shown in this reconstruction. The mastodon had a long, low body with a somewhat flattened skull, while the Columbian mammoth was taller with a sloping back and a very domed skull.

	Mammoth	**Mastodon**
Genus	*Mammuthus*	*Mammut*
Species	*M. columbi*	*M. americanum*
Height	4 m (13 ft)	2.4–3 m (8–10 ft)
Period	150,000–10,000 BP	1.6 million–10,000 BP
Location	North America as far south as Mexico	North and Central America

stripping their bark and leaves and occasionally pushing them over. In contrast, the slightly smaller woolly mammoth, with a shoulder height of up to 3.3 m (11 ft) had taken a different evolutionary route. It evolved in northern China and Siberia, in an area that experienced cool grasslands long before they became common in Europe. It adapted to eating lower-growing shrubs and dwarf willows as well as grasses. This type of herb-rich grassland has become known as the 'mammoth steppe' and it has no modern comparison. Therefore, in cold stages the straight-tusked elephant could not survive in northern Eurasia because the trees and shrubs that it relied upon for food had retreated south away from the cold, while for the mammoth the spread of the grasses associated with the cold steppe meant that the conditions in Europe were ideal in glacial periods, and less than perfect in interglacials – as forests replaced grasslands the mammoth food plants were lost.

Mammoths on the move. The reconstruction shows a small herd of mammoths moving through a snowy landscape. Woolly rhinos and horses are frequently found as fossils in the steppe, yet the lions shown feasting on the carcass of a reindeer are much rarer. Predators are always less common than prey animals (otherwise there would not be enough prey for predators to eat), hence the relative rarity of carnivore remains in comparison with those of the herbivores.

The mammoth steppe was not a cold snowy wasteland. Mammoths, like modern elephants, would have needed huge amounts of food every day in order to survive. This would require roaming over large distances to feed a herd, and if every mouthful had to be dug from beneath a blanket of snow mammoths would spend as much energy foraging as they would receive from the food. Hence it's more likely that they ranged through open grasslands with limited snow cover. The reconstruction shows life on the mammoth steppe 40,000 years ago. Typical cold-stage mammals of

Eurasia, such as the horse, woolly rhino and mammoth, are shown together in a mountainous landscape with grasses, small shrubs and trees in sheltered valleys. However, there is another element that is often overlooked – the piles of dung that would have been left by the mammoths as they traversed the grasses. Mammoths could have eaten up to 180 kg (400 lb) of plant food a day and such a huge amount had to go somewhere. This food has been found at every stage, from between the molars of a permafrost mammoth that died mid-chew, to partially digested foodstuffs in the stomachs and colons of mummified mammoths, to the final product of dung found in caves in arid parts of North America. From this material, at all stages of the cycle from mouth to steaming pile, the food that the mammoth was eating can be identified from the small pieces of plant stem and pollen that are preserved within it. These show that they ate a wide variety of foods, mainly grasses but also tree bark and leaves, as well as small shrubs.

Mammoths would have traversed the steppe in herds. Imagine a small calf at the back of the group with his mother. He would have followed the herd, feeding occasionally on plants, as shown by the wear on his teeth, but still mainly suckling milk from his mother. One day before his first birthday he had the misfortune to become stuck in either a mud pool or crevasse, and there he died. It is likely that there would have been bellowing and a lot of movement as the herd registered their distress at the loss of a member, and the presence of his mother

above These Columbian mammoth teeth from Hot Springs, South Dakota, show the characteristic plates of enamel that make up elephant teeth. The enamel ridges are held together with dentine, which is softer than enamel and wears away, leaving the enamel as ridges that cut through the plants as the elephant grinds its food. At any one time elephants have two molars in the lower jaw and two in the upper. As each tooth is worn away a new one erupts behind it; once the last one has erupted and worn away the animal dies as it can no longer chew its food.

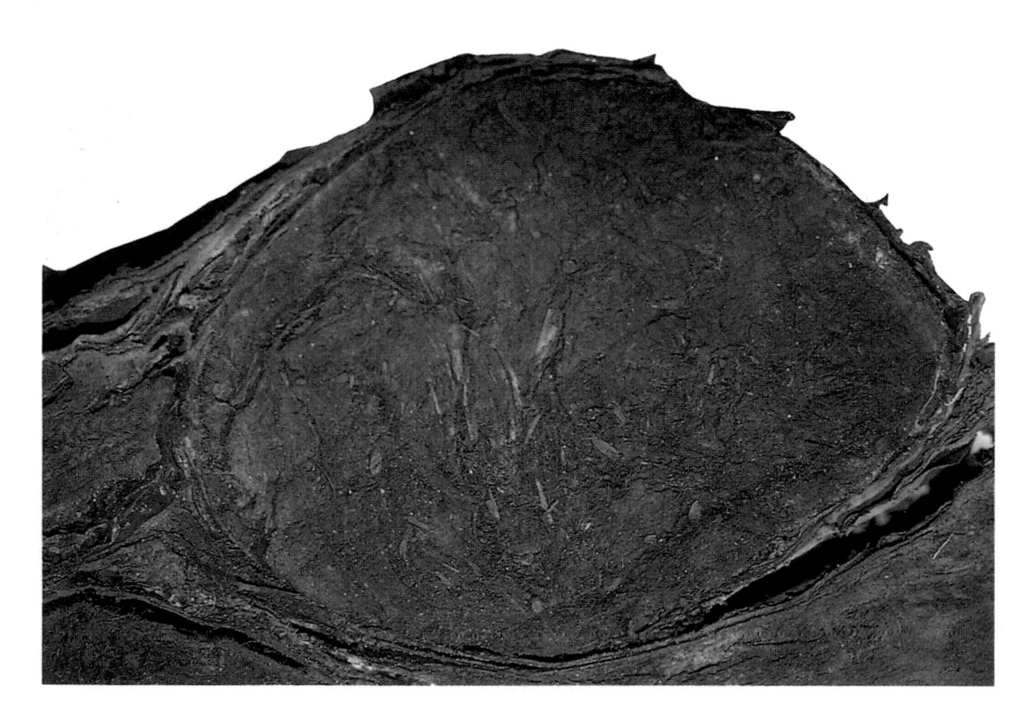

left A woolly mammoth may have eaten up to 180 kg (400 lb) of food per day. This photograph shows the arrested passage of food through the intestines of the Shandrin mammoth. A male mammoth found in permafrost in 1972, his guts were still packed with grass and leaves that he had consumed shortly before his death.

may have prevented predators and scavengers from feasting on him until he had drowned and become buried in the mud. And there he was discovered, almost complete, in the permafrost some 40,000 years later in 1977. He was named Dima after the tributary of the river on which he was found. Palaeontologists have pieced together the story of Dima's last days and hours from his body and the surrounding sediments. Dima was not a healthy calf. Analyses showed he had lots of parasites in his guts and had no fat deposits, indicating that he was malnourished. This may have weakened him so that he was unable to climb out of the hole in which he became stuck, but perhaps it was just too deep, as his mother would surely have tried, like modern African elephants today, to use her trunk to help him out. Dima's stomach contents also told their tale, containing little but soil and his own hair, perhaps chewed off in his distress and hunger when no longer able to reach milk or vegetation. This story is unlikely to be unique, and several mammoth calves have been found in the permafrost, the most recent and complete of which is Lyuba, discovered in May 2007.

Dima, a baby woolly mammoth, shown lying in the Siberian permafrost in 1977. Note the very small ears of these cold-adapted animals, in comparison with those of a modern African elephant calf. The sediment around Dima's corpse contained hair that had sloughed from the body, but the calf still retained hair on his trunk, tail and feet.

Frozen in Time

The majority of extinct animals are known simply from their bones and teeth, but just occasionally conditions are right for skin, hair and other tissues to survive. Frozen remains of horses, bison and mammoths, as well as snowshoe hares and ground squirrels, have been found in Siberia and Alaska, while sloths are the most frequent finds in the Americas, and complete Tasmanian tigers have been recovered in Australia. The permafrost finds are of animals that were quickly buried before scavengers and microbes could destroy the carcasses, while those from arid caves in the Americas and Australia are very dry, meaning that the decomposing microbes cannot get the moisture they need to survive and multiply. This produces two types of mummification; in the permafrost the animals are complete and may retain their shape but the hair is often sloughed off, while in cave sites it is the skin and hair that is best preserved, and the bodies are often flattened through desiccation. A third very unusual type of preservation has occurred at Starunia in what is now western Ukraine, where natural deposits of oils and salts have literally pickled the remains of woolly rhinoceroses, resulting in the most complete specimens known of this species. Naturally mummified remains of animals, particularly bats, can still be found in modern caves if the conditions are right.

opposite This baby mammoth was found in permafrost in northern Siberia in May 2007, and named Lyuba after the wife of her finder. Studies suggest that the calf was female and less than a year old when she died.

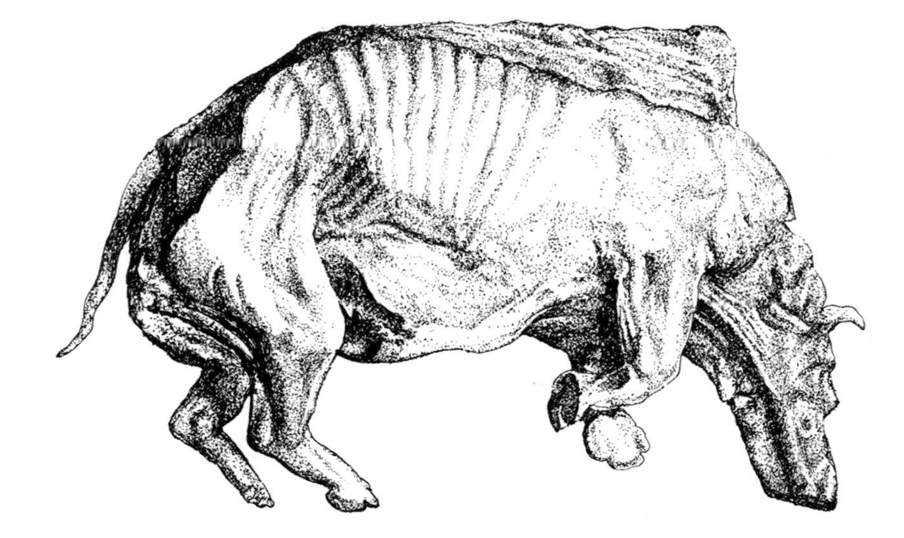

right The Starunia woolly rhinos found 'pickled' in oil and salt deposits in western Ukraine are some of the best examples of woolly rhino mummies that have yet been discovered.

below Exceptionally well-preserved animals have been found across the northern latitudes, as shown here.

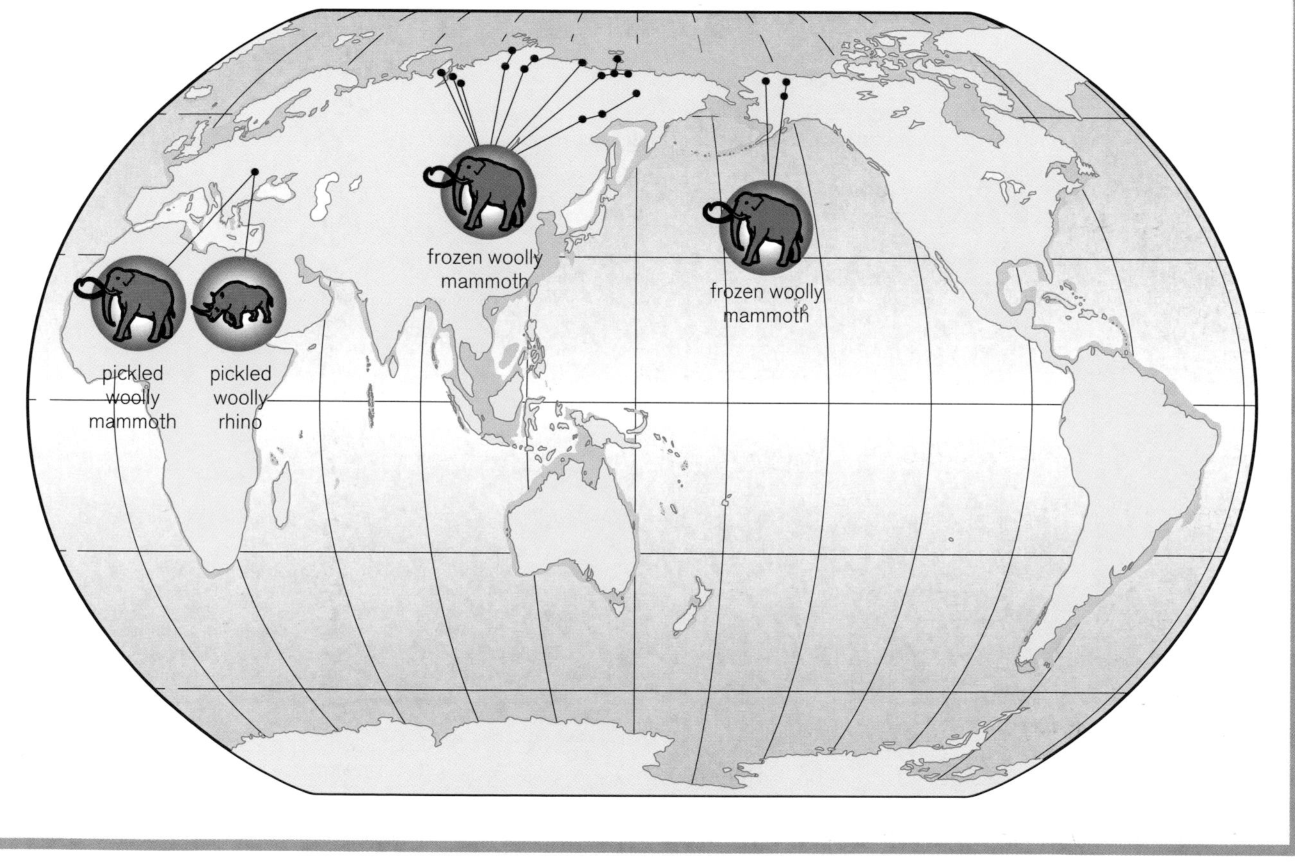

These frozen mammoth calves are often complete, in contrast to the corpses of larger animals, because their bodies are small and after death they would have quickly cooled down. The stomach contents of adults would have kept fermenting, even when the rest of the animal was frozen solid, leading to decomposition even if the animal was not scavenged. Hence few complete adult mammoths have been reported.

This engraving from Rouffignac in France shows many of the features of the woolly mammoth, with long, curving tusks, a furry coat (illustrated by the many downward strokes around the outline of the body), and a bulbous domed skull. The artist clearly knew his subject.

Today we are fascinated by mammoths because they are extinct, and the permafrost mummies give us a very clear idea of how they must have lived and died. Humans in Ice Age Europe must also have been impressed by them as they painted them on cave walls, along with woolly rhinos, lions, horses and bison. These animals must have been of importance, and they also commanded respect. The lions and hyaenas were larger than their modern relatives and they denned in caves, which people also wanted to use for shelter. This would have led to competition for caves, as well as for kills. Humans also used carnivore teeth for pendants and necklaces, which would have required hunting the foxes and other small animals from which the raw materials came.

The interglacial faunas from Europe provide a distinct contrast with those from the cold mammoth steppe. Britain in the last interglacial looked very different from anything we would find today, as spotted hyaenas, straight-tusked elephants, Merck's rhinoceroses, lions and hippopotamuses roamed the landscape. This would seem to be a land filled with animals, but two widespread mammals were missing, although they were present elsewhere on the continent – horses and humans. The lack of horses seems to have had quite a surprising effect on the spotted hyaenas. Horses are usually one of their favourite prey and they run them down over open grasslands. Britain lacked horses and was heavily forested, which meant that these hyaenas had to work much harder for their food, and when they got it, they did not waste it. Studies of their food remains show a consistent sequence of chewing in which they turned many of the long bones of animals even as large as a rhinoceros into bone cylinders (like napkin rings) by chewing off both ends.

Why were humans and horses missing during the last interglacial? It is likely that they did not reach mainland Britain before the North Sea and English Channel flooded after the end of the preceding glacial (125,000 years ago). There was a limited amount of time to reach Britain between the beginning of the thaw and the flooding of the Channel, and as animals were not moving purposefully but were just following vegetation as it grew, or in the case of the humans maybe following herds of other animals for hunting, the chance was there in each interglacial for some to be left behind. So in Britain humans and horses are missing from the last interglacial, and hippos are missing from the one before that. This also explains why there are no snakes in Ireland today, as the Irish Sea flooded before the Channel, so snakes reached Britain but not Ireland.

Given that we currently live in a warm phase, why do we recognize the cold-stage animals and know far less about the warm-stage creatures? There was a glacial immediately before the warm phase in which we now live, so the most recent remains of extinct animals are all from a cold period and their bones abound in both glacial deposits and in archaeological sites. Modern humans arrived in Europe during a cold stage about 40,000 years ago and they preyed on, and lived alongside, these amazing creatures. The engravings, carvings and vivid cave paintings that they produced provide us with magnificent portraits of the long-extinct animals that they lived among. The cold-stage animals are also those that wandered the high tundra in northern Siberia and America, some parts of which have remained frozen for thousands of years. The Alaskan and Siberian permafrost occasionally yields the remains of mammoths (like Dima), bison, horses and even creatures as small as ground squirrels. In contrast we have only the bones and teeth of warm-stage species. They come to light throughout Europe, but have survived, often much

overleaf These paintings of Ice Age beasts, including rhinoceroses, wild cattle and horses, predate the Lascaux bas-reliefs by some 10,000 years. This spectacular rock art was found when the site of Chauvet Cave was discovered in 1994. It is likely that most paintings within the cave were created between 32,000 and 30,000 years ago, although people revisited the site 27,000–26,000 years ago.

below Upper Palaeolithic bas-relief carvings of aurochs (wild cattle) from the site of Lascaux in France. Estimated to be 16,000 years old, these carvings were created by modern humans towards the end of the last glacial period.

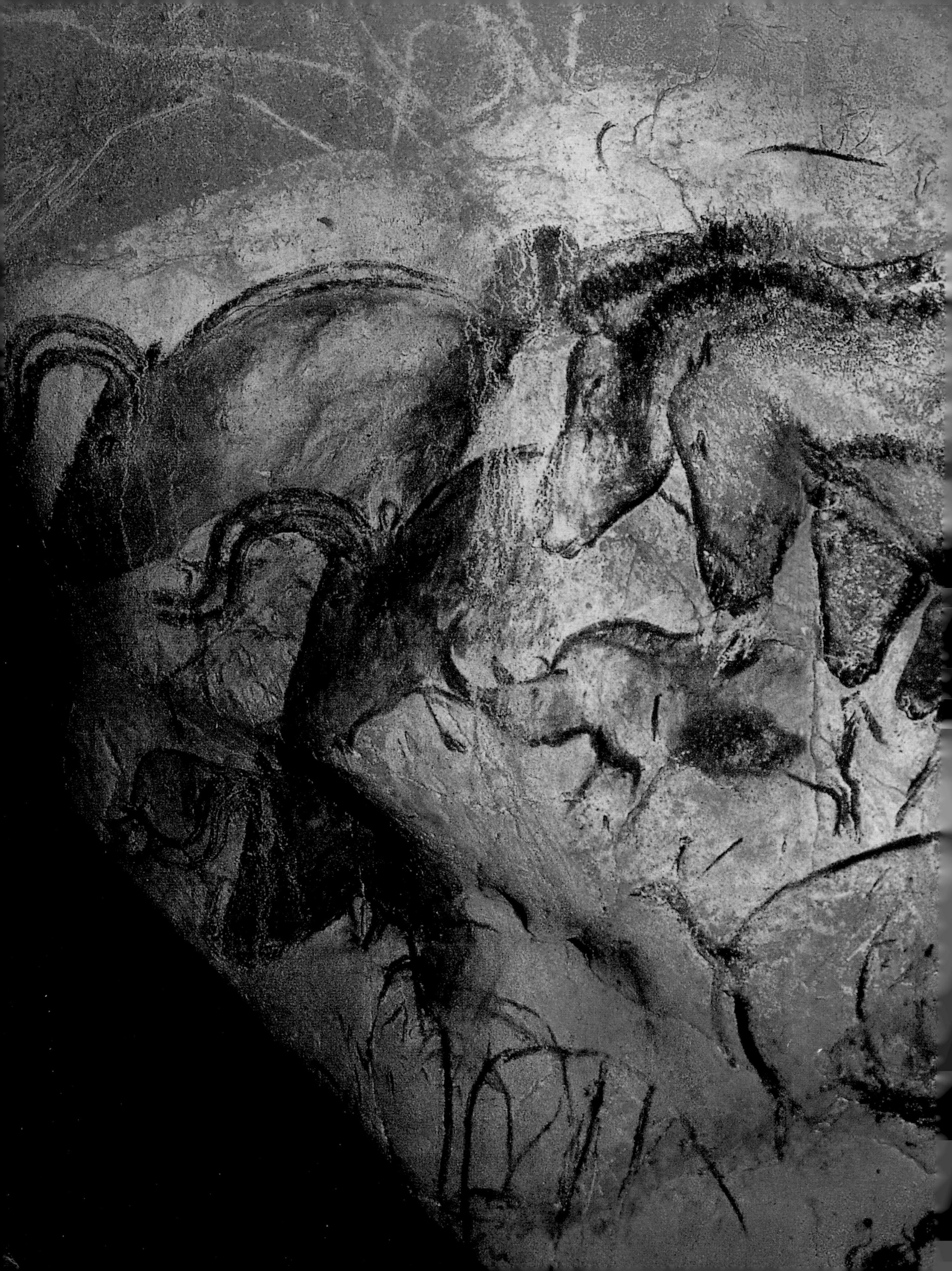

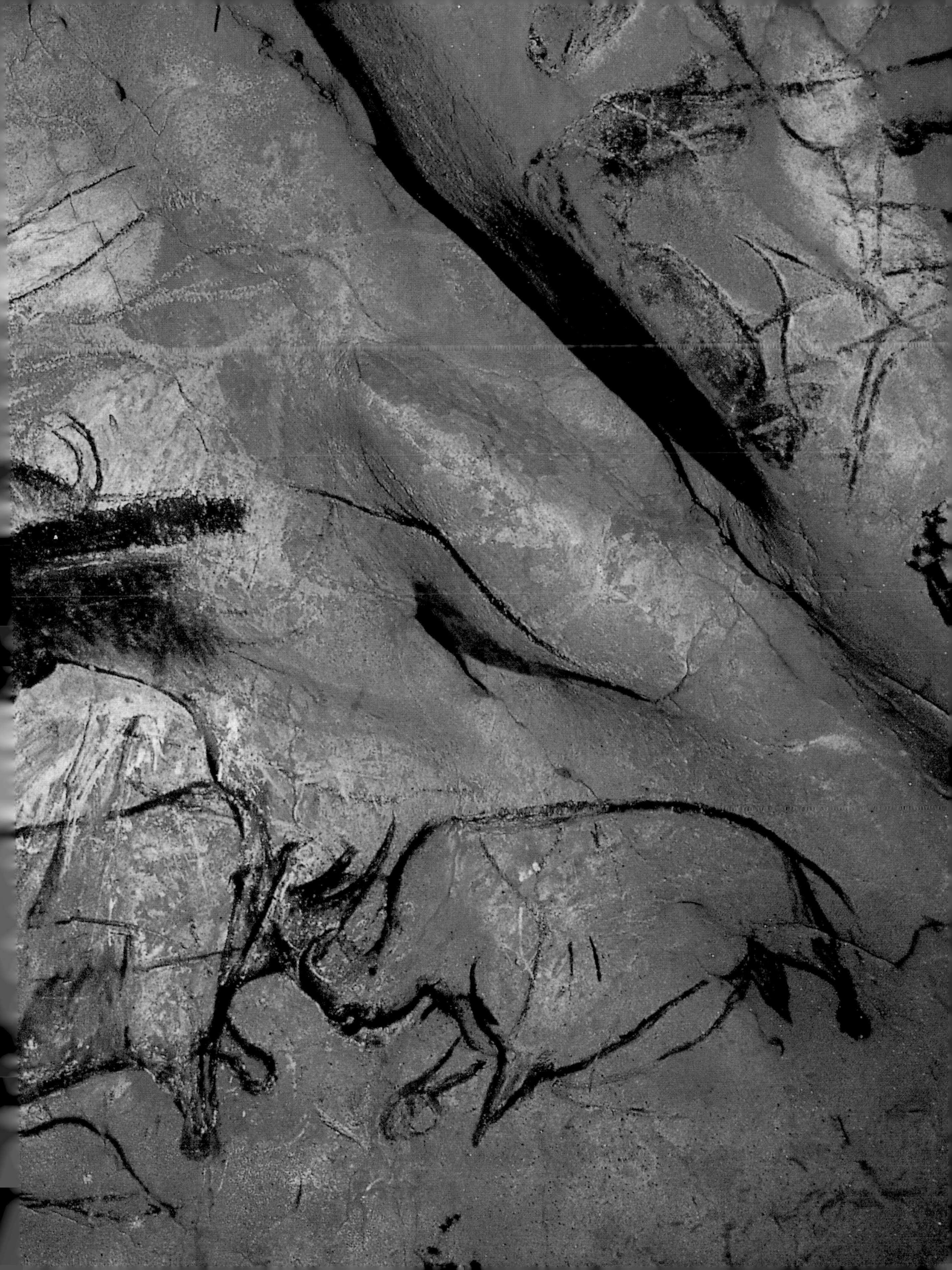

eroded, through at least one glacial period. In some places, such as Britain, the evidence has been all but obliterated by the advancing glaciers, or deeply buried by huge quantities of sediment deposited by glaciers and meltwater rivers. The transport of such enormous amounts of soil and gravel explains why so many fossils are found only by chance during quarrying and mining. Pockets of fossils also survive in other places such as caves – without the chance survival of last warm-phase fossils at Victoria Cave in northern England we would not have known that hippopotamuses were living in the Yorkshire Dales 120,000 years ago.

Many of the animals that inhabited Europe in the interglacials of the last 500,000 years are only found in Africa in modern times, or are now completely extinct. We ourselves are living in an interglacial, but one that is unusual because these species are missing. We are left with the sad remnants of a once-incredible fauna, and it was not just Europe that suffered extinction – in fact the European fauna survived relatively well compared to that of the Americas and Australia.

above These great holes in the ground are the footprints of *Megatherium*, the giant ground sloth. Walking in soft sediments, the vast bulk of the animal caused the ground surface to sink, creating the dents in which water can be seen pooling in the photograph. Footprints provide a rare insight into the movement of extinct creatures, as the gait, speed and weight of the animals can be calculated from their tracks.

North America

If you could step back into North America 20,000 years ago you would find yourself surrounded by camels, four types of elephant (woolly mammoth, Columbian mammoth, mastodon and gomphothere), jaguars, American lions and pumas. And that's not all, as ambling among those readily identifiable creatures would have been others for which we have no modern comparison – giant ground sloths over 3 m (10 ft) high, armadillo relatives the size of cars, an enormous hunting bear over 1.6 m (5.5 ft) tall at the shoulder and a sabre-toothed cat that may have been capable of catching baby mammoths. These animals formed a community that is hard to imagine today, but they were not isolated. They could move out of North America in two directions, across Beringia into northern Asia (at times of low sea level), and across the Isthmus of Panama into South America. Animals could also cross in the opposite direction, and it appears that that is what they did, with woolly mammoths and bison arriving across the Bering land bridge at some point in the last 100,000 years. Curiously, the woolly rhinoceros, although present in northern Siberia, did not make it across to North America. Perhaps the most famous traversers of Beringia were the humans who crossed towards the end of the last glacial.

The fauna in North America varied across the continent, just like today, when animals living in Florida are different from those in Washington State or Arizona. Some creatures, such as Rusconi's ground sloth, were at the northern extent of their range, having spread up from South America through Mexico to just reach into what is now the United States, while the most southerly finding of the musk ox is in California. Vegetation also played a part in the distribution of animals, as out on the

The giant ground sloths of the genus *Megatherium* were found throughout South America and as far north as Texas. The largest of all the ground sloths, they could weigh up to 2,700 kg (5,950 lb), but despite their bulk they were probably capable of squatting on their hind feet, as shown here. The front claws were used to pull down branches from trees to eat their leaves.

Megatherium	
Genus	*Megatherium*
Species	*Megatherium americanum*
Height	2.1 m (7 ft)
Period	4 million – 10,000 BP
Location	South and Central America

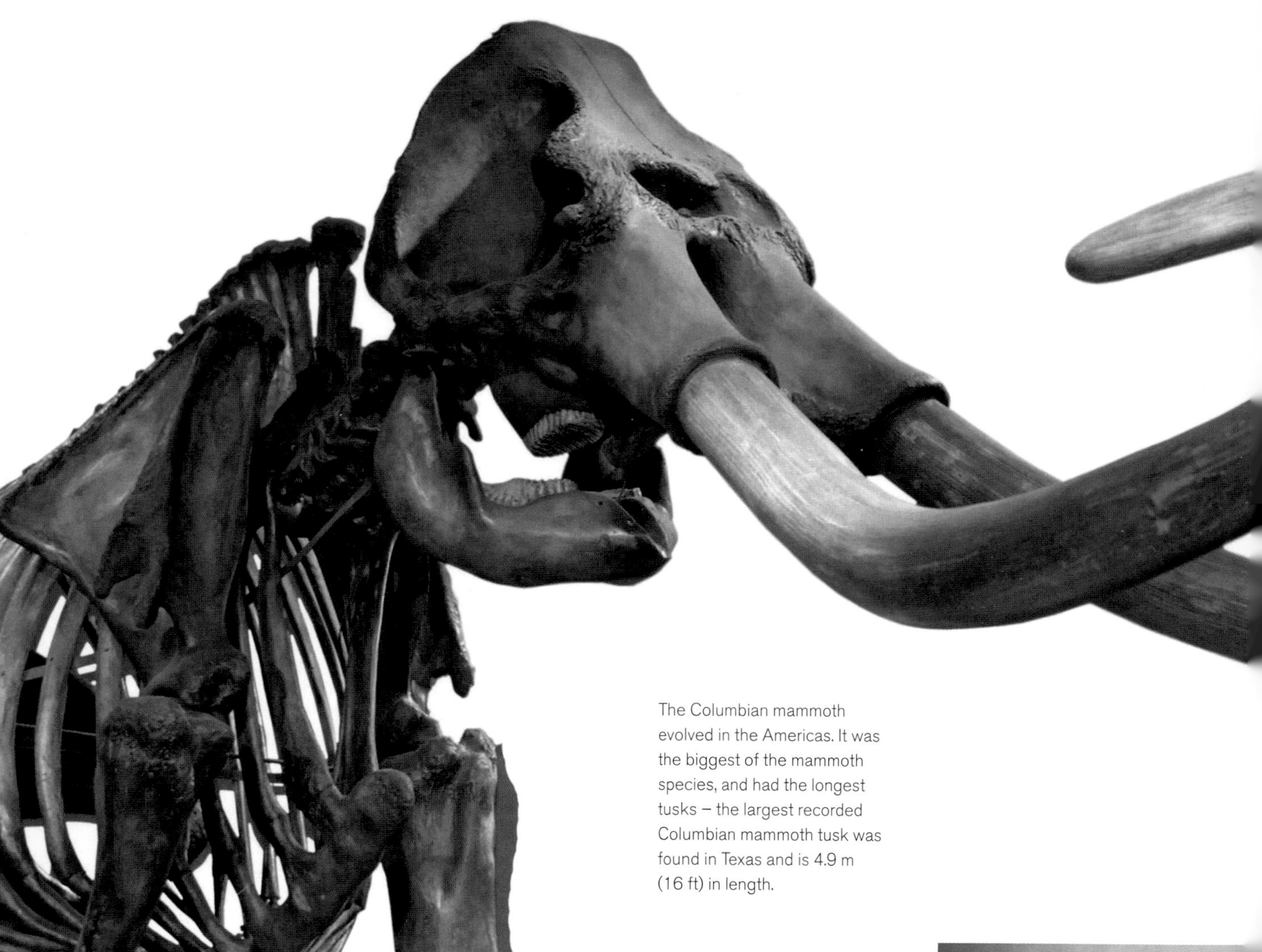

The Columbian mammoth evolved in the Americas. It was the biggest of the mammoth species, and had the longest tusks – the largest recorded Columbian mammoth tusk was found in Texas and is 4.9 m (16 ft) in length.

plains among the pronghorn antelopes, herds of Harlan's ground sloths, 3.6 m (12 ft) tall and weighing 1,500 kg (3,300 lb) would have been grazing slowly forwards, while yesterday's camel would have been browsing at the forest edges. The Florida bear was found nowhere else but North America, and the elephant-like mastodons and gomphotheres were the last surviving representatives of families that had flourished throughout the world in earlier time periods, but now had a final stronghold in the Americas. Mammoths dispersed twice into North America: the ancestors of the Columbian mammoth arrived first and this species evolved *in situ*, while towards the end of the Ice Age the woolly mammoth came across the Bering land bridge. The Columbian mammoth was the largest of all mammoths, with a shoulder height of 4 m (13 ft). Given that it lived in warmer climates, although sometimes at high

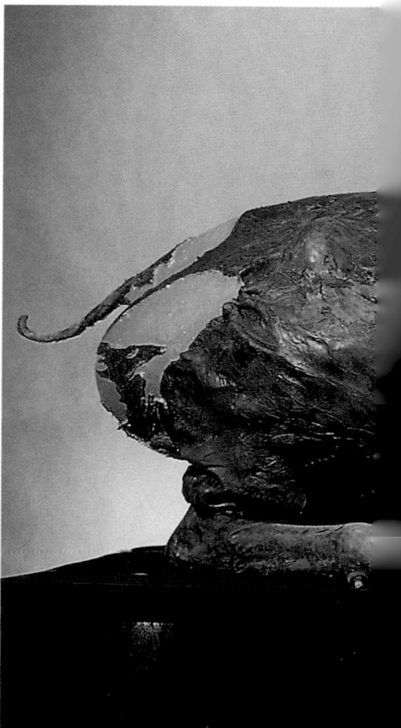

altitude, it is likely that it was not as hairy as its arctic cousin. Unlike the mammoths, some other animals were found in restricted areas, with many found in Central and South America only reaching as far as the southeast portion of the United States, around Florida, where the truly giant ground sloths and gomphotheres trod.

In the north lived the mammoth steppe fauna that stretched from Europe across Beringia and into the Americas. Life is tough for all animals, a constant battle to find food, find a mate and stay safe. Dima's story (above) highlights the role of chance in survival, but accidental death, old age or diseases are not the only things that can kill an animal. The story of Blue Babe, a bison in the prime of life, gives us insight into another aspect of life on the mammoth steppe – the battle between predator and prey. Blue Babe was a steppe bison, aged between 8 and 9 years, who, to judge from the extensive fat deposits on his body, had spent a good year grazing on the Alaskan

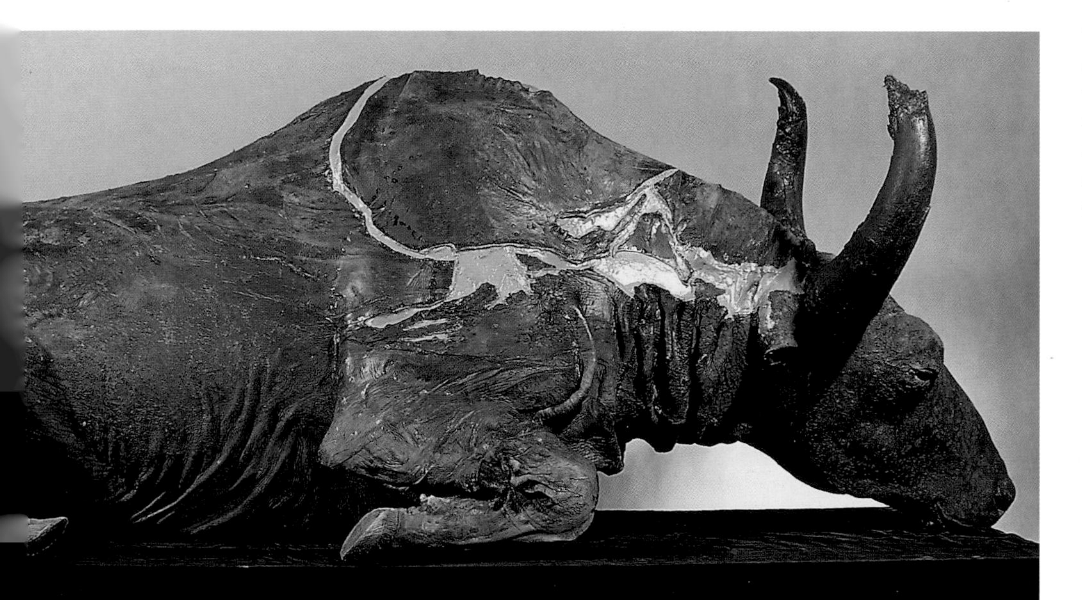

The stuffed skin of Blue Babe, a male steppe bison, *Bison priscus*, found in the Alaskan permafrost in 1979. Many of the holes and scratchmarks on the skin were made by lions when attacking and later feeding on the carcass.

steppe. One day in early winter 35,000 years ago, he had the misfortune to attract the attention of a small group of lions. Male steppe bison probably lived alone, so there would have been no other bison maintaining vigilance to warn Babe of the lions' approach. Alone and grazing in a small river valley he was attacked from behind, with the lions using their paws to try to knock him off balance, as shown by the numerous claw marks around his back and legs. Eventually one of the lions managed a muzzle bite (when a lion clamps its jaws over the mouth of its prey to suffocate it) and the bison was killed. The lions then opened the carcass along its back and devoured the muscle and bones. However, this feast would only have lasted a few days until the corpse froze solid in the cold Alaskan winter, following which the carcass was abandoned – although one lion did have a last attempt to eat the meat, and left part of its tooth lodged in Babe's now-frozen skin. This drama would have been played out thousands of times on the steppe, but Blue Babe is special because he became an Ice Age mummy, so that many of the details normally lost when an animal decays were preserved for scientists to study. He was found during mining activity in 1979 and examined in detail by palaeontologist Dale Guthrie. The researchers did not just stop at studying him; apparently without ill effects they ate part of his neck in a stew to celebrate the end of the research project. The stuffed skin is now on display at the University of Alaska Museum in Fairbanks.

The tale of Blue Babe highlights the competition between predators and prey. In the Americas there were lots of different prey species, including sloths, mountain goats and pronghorns, and there was a wealth of large carnivores seeking to consume them. One strategy to avoid being eaten is to be very large – so big that even the largest mammalian carnivores cannot successfully attack you. This is the strategy followed by modern elephants, and also by rhinos and hippos. North America had four types of elephant, but also huge ground sloths, including Rusconi's ground sloth, which had a range from Texas to South Carolina. This sloth was the biggest to inhabit North America in the Pleistocene (although there were even larger ones in the south), and was up to 6 m (19 ft) long, weighing in at 2,700 kg (6,000 lb). Although a peaceable vegetarian, this animal was armed with claws, and those, combined with its sheer size, would have made an adult invulnerable to predators. However, the young of large mammals may have been in more danger. A site in Texas named Friesenhahn Cave contains the bones of many juvenile mammoths and sabre-toothed cat cubs. It has been suggested that this cat, *Homotherium*, specialized in preying upon juvenile mammoths at about 2–4 years of age when they would have been adventurous and leaving their mothers for short periods. No modern animal specializes in hunting baby elephants, but perhaps *Homotherium* did.

left *Glyptodons* were slow-moving but had excellent natural defences, as they were covered with a carapace made up of many bony plates. However, they were not without enemies – there is evidence that they fought between themselves (below), and occasionally made a meal for large carnivores. They were found mainly in South America, but did spread up into what is now the southern United States.

below The armoured tail of a *Glyptodon*. *Glyptodon* tails were well defended and some even had clubs on the end. It is thought they were used for protection, defence and possibly for fighting other glyptodons for access to food or mates.

Another popular anti-predator technique is to grow armour, like the modern armadillos. The armoured animals of the Ice Age ranged in size from fairy armadillos up to 15 cm (6 inches) long to truly enormous glyptodons that were 3 m (10 ft) long. Some of the smaller animals would have been capable of rolling up to defend themselves, but others had a tortoise-like shell made up of a mosaic of irregularly-shaped bony plates. These glyptodons could have browsed in peace, snug within their home-grown armour, but they did have violence in their lives. As well as a hard shell, many species had a fused collection of bones at the ends of their tails that formed a club. This weapon could have been used for defence against predators, but the finding of shells with club-sized dents in them has also suggested that glyptodons also used their tails to fight each other, perhaps for access to females or food.

This picture of *Smilodon* illustrates many of the features of a sabre-toothed cat. The sabre-tooths had a powerful build and robust forelimbs to tackle their prey and wrestle them to the ground, and the tremendously elongated canines that give them their name can be readily seen in this reconstruction.

What were the predators that may have threatened these animals? North America had a huge array of large carnivores, only comparable today with the plains of Africa. There were two sabre-toothed cats, *Homotherium* and the infamous *Smilodon*, found by the hundred in the tar pits of Rancho la Brea in California, where they had been attracted by the smell of animals mired within the sticky asphalt. *Smilodon*, like its European ancestors, was probably an ambush hunter, hiding behind trees and rocks and pouncing on animals as they went past. With flattened canines 17 cm (7 inches) long it must have had to time its attack carefully to avoid breaking them, as there was no chance of re-growing them once they were broken. Exactly how sabre-tooths attacked and killed their prey is still a subject of debate among palaeontologists. Some have suggested that they used their canines to slice open an animal's belly and then waited for it to die of shock or blood loss, while others

Genus	*Homotherium*	*Smilodon*
Species	*H. latidens*	*S. fatalis*
Shoulder height	1.1 m (3.4 ft)	1 m (3 ft)
Period	3 million–10,000 BP	1.6 million–10,000 BP
Location	Africa, Europe and Asia	North and South America

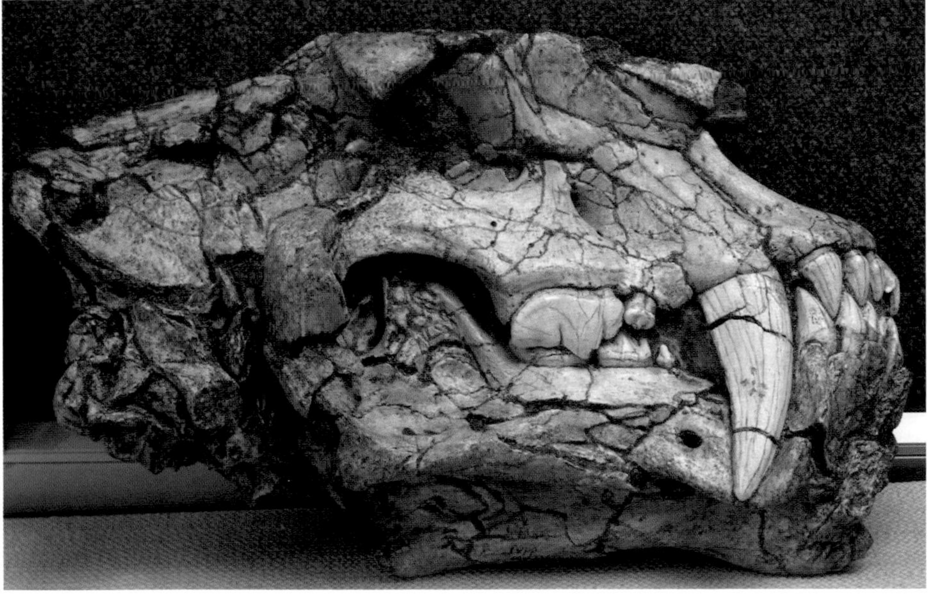

above Thousands of *Smilodon fatalis* bones have been found at the site of Rancho la Brea tar pits in California. Many of these bones have pathologies, indicating that life was hard for the top predators, or that the injured cats were driven to scavenge off already dead and dying creatures.

left Several genera of sabre-toothed cats existed during the Ice Age in different parts of the world. This skull of *Homotherium* from China shows the large slicing cheek teeth, the rounded arcade of incisors used to tear meat from the bones of their prey, and the slashing canines, all typical sabre-tooth features.

The dire wolf, *Canis dirus*, was found in North and South America. More robust than the modern grey wolf, it may have been an opportunistic scavenger. Its remains have been found in great numbers at the tar pits of Rancho la Brea in California.

think that they used their robust forelegs to wrestle with an animal and once immobilized the sabres were used to cut the jugular. No living animal has these adaptations, so we cannot know for sure. The other cats are more familiar to us because all bar one are still alive today. These were the American lion, a very large version of the modern African species, the jaguar, puma and the American cheetah, a distant relative of the modern cheetah, but somewhat larger. Hyaenas became extinct halfway through the Ice Age in North America, but the dire wolf, a larger, chunkier version of the grey wolf, and a huge short-faced bear may have taken their roles. The short-faced bear may have been an aggressive scavenger and it would have been able to defend kills from any other single carnivore, although it may have been overwhelmed by group hunters like grey wolves.

One animal about which we know an enormous amount was one of the smallest ground sloths, the medium-sized Shasta ground sloth that lived throughout North America from Canada to Arizona. This creature weighed 135–545 kg (300–1,200 lb) and left an incredible legacy in the caves of the southern United States – tonnes and tonnes of dung. One site, Rampart Cave in Arizona, contained an estimated 220 sq. m (2,370 sq. ft) of sloth dung. This treasure trove of information has been pored over by scientists to extract the data held within it. Like mammoth dung it contains plant fragments and pollen that show the sloth was a vegetarian, visiting the site in winter and early spring to eat globemallow and opuntia cacti. These plant remains have been radiocarbon dated and the results showed that the sloths had made the same trek to the site for at least 30,000 years before dying out at the end of the last glacial. However, diets and dates are not the end of the dungy delights, for the cave also contained the remains of four new species of stomach parasites, including worms which may have been unique to the Shasta ground sloth.

Nearly all these large mammals became extinct at the end of the last glacial (see below). The Shasta ground sloths' journeys in Arizona came to an end, the forests were no longer browsed by mastodons and the glyptodons left the plains. The disappearance of these large herbivores had a massive impact on the animals that preyed upon them; following in their wake the sabre-toothed cats, giant bears and dire wolves became extinct. Four different members of the elephant family also vanished, leaving the Americas without an elephant for the first time in millions of years. This left just those animals with which we are familiar – the pronghorns, armadillos and bison – to enter into the current warm phase, being preyed upon by pumas, jaguars, grey wolves and humans.

Australia – experiments in evolution

Australia was originally a forested continent, which has become progressively drier and more open over the past 15 million years. The desert dunes that are present today appeared only 1 million years ago, which means that the Australian fauna has had to cope with enormous climate change, and with few links to other areas to bring in new species. In fact, until humans appeared, the only terrestrial animals to arrive since Australia's isolation by sea 35 million years ago were rats and mice which have made the journey three times in the last 5 million years.

The relative isolation of Australia had benefits for a type of animal that had become extinct elsewhere in the world; nowhere but Australia and New Guinea did the monotremes (echidnas and platypuses) survive. These are often called 'primitive' mammals, because they lay eggs while all others grow their embryos inside them. However, the eggs quickly hatch and dependent young are looked after for some months, sometimes in pouches like marsupials. Monotremes are one of only two groups of poisonous mammals (the other consists of some shrews). The male platypus can stab its attackers with a spur that delivers enough venom to kill an animal as large as a dog. There are

The duck-billed platypus, *Ornithorhynchus anatinus*, has several claims to fame: it is one of the few egg-laying mammals in the world, and is also one of the few poisonous mammals. It is only found in eastern Australia, where it lives in burrows in river banks and hunts underwater for its insect prey.

four living species of echidna, all of which have spines on their backs and lick up ants, worms and other small prey with their long sticky tongues. A very large echidna, perhaps 1 m (3 ft) long, is known from fossils from western Australia, but so far the head of this animal has not been found.

The Australian mammals look strange to our eyes – pouched (marsupial) beasts and egg-laying hedgehog look-a-likes – yet relatives of these animals were once found on other continents, and only the evolution and dispersal of the placental mammals precipitated their loss. When the first humans arrived in Australia they would have seen multitudes of hopping and browsing marsupials that had evolved to fill all sorts of niches from burrowers to predators. The only other place that such marsupial variation occurred in was South America, but here most became extinct after the Great American Interchange. Therefore, in the living and recently extinct faunas of Australia we get a glimpse of the past before the marsupials had been outcompeted in the rest of the world. This is not to say that the modern Australian animals have not evolved – they certainly have! Adult platypuses now have just a horny beak to chomp their food, but fossils from 10 million years ago show that they

The Komodo dragon, *Varanus komodoensis*, is the world's largest living lizard, and its habits probably resemble the extinct giant Australian lizard *Megalania*.

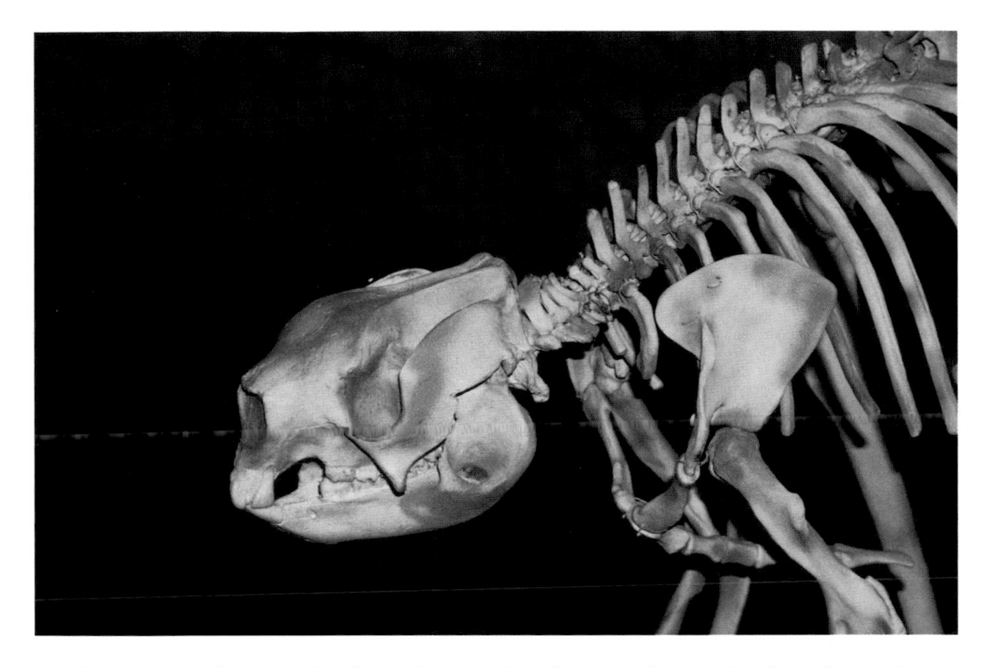

There were many different marsupials in the Ice Age. This is a skeleton of *Sthenurus tindalei*, an extinct wallaby-like animal, whose remains were found in the Victoria caves near Naracoorte in Australia.

used to have teeth. Giant koalas twice the size of the modern animal used to sit in the branches of the eucalyptus trees, and in the last few hundred thousand years grey kangaroos and wombats have evolved rapidly into many different species to take advantage of the spreading grasslands. These examples illustrate that Australian mammals have not been in stasis; they have changed through time, but they could not alter their basic biology – they were monotremes or marsupials and that was all the material evolution had to work with.

Unlike all of the other inhabited continents, the largest predators in Australia were not sabre-toothed cats or bears, but giant reptiles – a giant monitor lizard named *Megalania* and a snake called *Wonambi*. *Megalania* was the length of the largest modern saltwater crocodiles (up to 7 m (22.5 ft) long depending on estimated tail length) and maybe weighed the same as a small rhino – 1,900 kg (4,200 lb). Fossils of this formidable beast have been found in north, central and eastern Australia in cave and river deposits, but they are always rare. *Megalania* was most closely related to the monitor lizards, and its largest living relative is the Komodo dragon, which bites its prey with a bacteria-filled mouth and waits for them to die from blood poisoning. Komodo dragons (and probably *Megalanias*) are also capable of ambushing and chasing prey over short distances. The giant monitor lizard's teeth were sharp and serrated to aid in cutting and it would have preyed on anything it could get its jaws on, although the only direct evidence of its diet is a partially digested wombat tooth found with some *Megalania* ribs. The second fearsome reptile was *Wonambi*, a large snake that would have constricted and crushed its prey to death. It was up to 5 m (16 ft) long and was the last surviving

member of a family that had lived among the dinosaurs but had since died out on all other continents. There were also two huge browsing tortoises, both of which were heavily armoured, with horns around the backs of their skulls and a large club on the ends of their tails. The dominance of reptiles as top carnivores in Australia may be a reflection of how these animals get energy from their food. Reptiles use fewer calories than mammals, so where food was scarce only a few good kills were needed to keep a *Megalania* fed for a year, whereas mammals need to eat much more frequently to stay alive.

If the reptiles were unusual, the mammalian predators were even stranger to our eyes. As well as the grass-eating lineages of kangaroos, there was another that had

above left A cast of the skeleton of the giant wallaby *Protemnodon*. Some species of *Protemnodon* weighed up to 100 kg (220 lb). The skeleton clearly shows why kangaroos are sometimes termed 'macropods' (literally 'big feet'), as they have a large but narrow hind foot, with only two toes (the fourth and fifth) taking the weight of the animal.

left The thylacine, *Thylacinus cynocephalus,* probably had a similar role to a small canid such as the red fox. This photograph, however, shows how the thylacine differed in bodily proportions, with a very long, thin tail, short legs and a relatively small head. It also had a striped coat, which led to its alternative name of the 'Tasmanian tiger'.

below For over 100 years bounties were paid for killing thylacines as they preyed upon livestock, but it is likely to have been a combination of persecution and disease that ultimately led to their extinction.

evolved to become an omnivorous scavenger. These creatures, named *Propleopus*, may have weighed 70 kg (154 lb) and, rather like a modern dingo, they would have eaten whatever they could find. Another predator was the Tasmanian tiger or thylacine. These dog-sized carnivores were extensively hunted during the European colonization of Australia and the last one died in 1936. Unlike the other extinct animals discussed we have a good idea of the colour, size and shape of the thylacines from photographs and films of the last few survivors and

the dried remains of thylacines that have been found in caves in western Australia.

Both *Propleopus* and the thylacine would have taken advantage of any food that they found, but one mammal had adapted to pure carnivory. This was the marsupial lion – a creature with scissor-like cheek teeth up to 4 cm (1.5 inches) in length. It had large claws on its front feet and may also have been capable of climbing trees. The marsupial lion was capable of slicing the flesh from its victims, but its teeth were so specialized for this role that it was incapable of crushing bones, limiting the amount of use it could make of a carcass and providing plenty of opportunities for scavengers. Even *Megalania* may have scavenged from its kills. The marsupial lion was the Australian equivalent of the sabre-toothed cats that we have seen in the Americas, a hyper carnivore so evolved that meat cutting was the only option.

So these were the carnivores, but what did they prey upon? The Australian marsupials appear to have evolved into three main shapes: small insectivorous animals that hunted for prey in crevices and up trees, larger hopping creatures that were insectivorous or plant eating, and the benign-looking wombat-types that ate vegetation. These wombat-types evolved into large-bodied creatures able to cope with eating and digesting large quantities of relatively poor food (consider the calories in a chocolate bar versus a salad and think how much salad you would have to eat to get the same amount of energy). But Australia was not a land of giants and, in contrast to the other continents, nothing the size of an elephant browsed its way through the Ice Age. The biggest mammals were the *Diprotodons* (meaning 'two front teeth'), which had beaver-like incisors but were much, much larger. At least four species of *Diprotodon* are known from the Pleistocene, one of which was the largest marsupial ever and the first fossil species

opposite The marsupial lion, *Thylacoleo carnifex*, was the largest marsupial predator in Australia, standing about 70 cm (27 inches) high at the shoulder. The large flesh-slicing cheek teeth can clearly been seen in the skull of this skeleton.

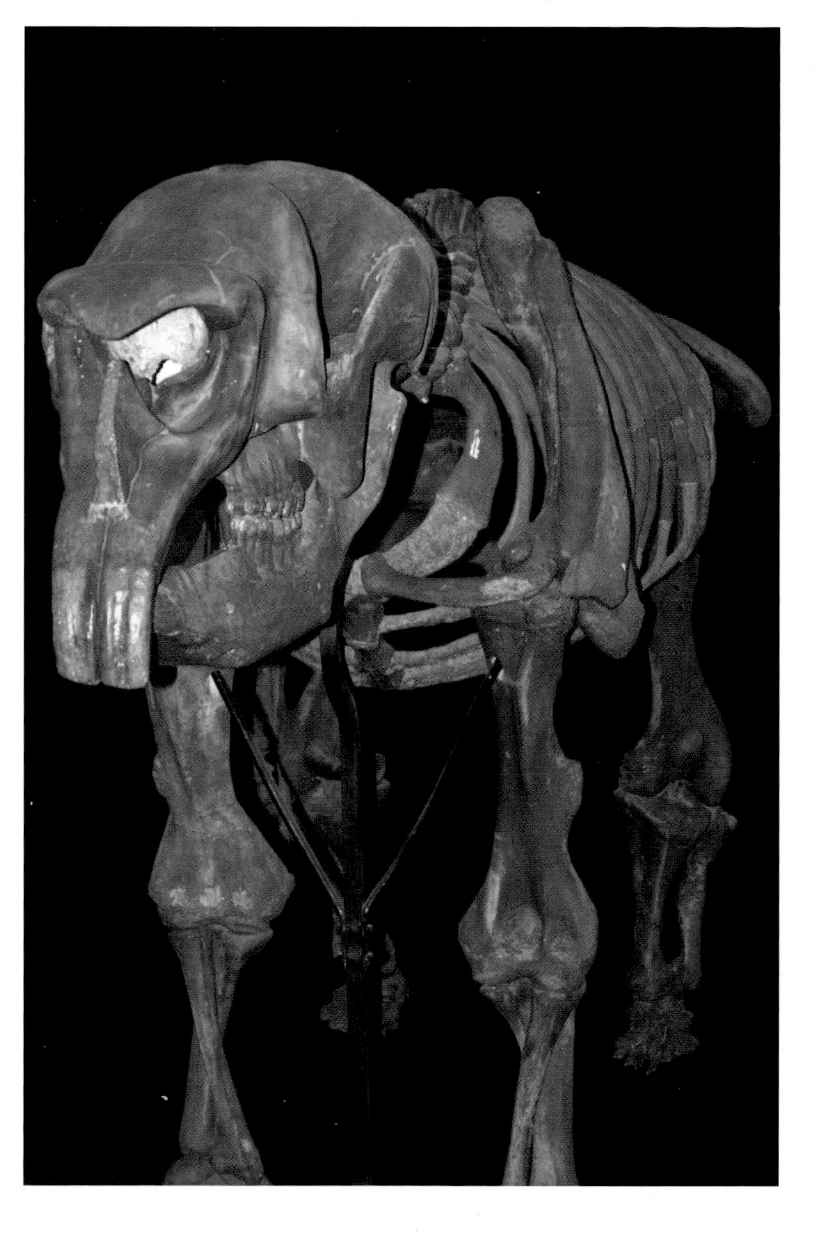

below *Diprotodon* was the first fossil mammal to be scientifically described from Australia in 1838. The beast was named *Diprotodon* (literally 'two first teeth') by Professor Richard Owen.

There were several species of *Diprotodon*, all of which are now extinct. *D. optatum* weighed up to 2,000 kg (4,400 lb), making it Australia's largest land mammal. Fur imprints preserved in soft sediments indicate that it had a hairy coat, as shown here on the reconstruction.

to be described from Australia. This browsing beast was 3 m (10 ft) long and 2 m (6.5 ft) high at the shoulder and inhabited parts of central Australia. In eastern Australia *Palorchestes* browsed. This animal, shaped rather like a ground sloth with large claws on its front and back paws, may have pulled down branches or dug up tubers using its front feet. The front legs were much longer than the back legs, and the nostrils were at the end of a short trunk, giving the appearance of a tapir grafted onto a ground sloth. Elsewhere, in forests surrounding waterways, the marsupial rhino, so called because it may have had a horn on its nose, lurked like a robustly built cow, seeking the plant foods on which it preyed.

These are just a few examples of the creatures that inhabited Australia, but by the end of the last glacial nearly all of them were gone. Thirty-five million years of evolution in isolation had come to an end. *Diprotodons*, *Palorchestes*, the marsupial lion and the large reptiles became extinct, the kangaroo and wombat families were drastically reduced and only the monotremes survived unscathed.

The demise of the Ice Age beasts

In each of the areas described, the mammalian fauna was varied and long-lasting. Yet today we have comparatively few species left – in the space of just 40,000 years, up until the end of the last glacial, the world lost 65 per cent of its large mammals. Now only two elephants remain where 100,000 years ago there were at least eight species. Not all continents were equally affected: Africa and Europe lost fewer than 10 groups, while South America fared worst, losing at least 50 of its large mammals. Many explanations have been put forward to explain why the fauna we have now is so impoverished, and why the majority of species disappeared within a very short period of time. These centre on two major incidents in the last 50,000 years – the expansion of modern humans into Australia and North and South America, and the rapid oscillations in temperature that occurred at the end of the last glaciation.

Modern humans took with them new technologies and would have entered lands that contained animals that were naïve to bipedal hunters. Predator naivety was famously described by Charles Darwin, who arrived on the Galapagos Islands and pushed a hawk off a branch using the barrel of his gun, while one of his companions killed a small bird with his hat. Running away requires energy, and animals such as mammoths and mastodons or giant kangaroos and *Diprotodons* would not have been used to predation of any sort, and would therefore not have learned to run away. In addition to stone tool technology, with modern humans came fire, and in some places such as Australia this may have had a massive impact on the plants. The general effect of fire on vegetation worldwide has been to suppress forests and promote grasses, amplifying the trends already introduced by climate change during the Ice Age.

It was generally the large mammals such as mammoths, ground sloths and *Diprotodons* that became extinct at the end of the last glacial – mice and other small mammals tended to be left unscathed. The large mammals take longer to mature and be able to reproduce, which means that they are more vulnerable to hunting, especially if those hunters concentrate on particular age groups such as juveniles or young adults. Not all individuals would have had to be killed for humans to have had an impact, as populations already made vulnerable through climate change, with resulting aridification and habitat changes, may have been small, and the arrival of hungry humans may have been the tipping point.

Alternatively, rapid changes in climate over a period of tens to hundreds of years may have altered vegetation so much that environments became unstable and large-bodied animals proved unable to find enough food to feed a viable population. Animals also disappeared for different reasons. Reindeer disappeared from France because their preferred habitat shifted to the north as the climate warmed, but the loss of warm-adapted mammals that should have benefited from the warming, like ground sloths and mastodon from North America, is more difficult to explain.

This is a controversial area, but one fact remains, that the megafauna survived the previous glacials, but not the last, and the thing that differentiates the last one from the others is the arrival of modern humans on all continents. Animals that had survived the changing conditions as small populations in the past could have been vulnerable to hunting within their refugia, perhaps precipitating their extinction. Large mammals such as the straight-tusked elephant have been referred to as 'keystone species', meaning that they provide the ecological structure that other animals can lean on. A browsing elephant may have made clearings in forests for other plants to grow, and it would have produced dung that was lived in and on by insects and fungi. Its intestines would have been teeming with bacteria and parasites, and its skin probably carried ticks and fleas, some of which may only have been found on that species. So the extinction of just this one animal would have had repercussions throughout the ecosystem – many other organisms would have lost their homes or food sources.

The surviving animals and plants had co-existed and evolved with the extinct animals over thousands of years. Considering these possible interactions can explain some of the otherwise puzzling characteristics of the surviving flora and fauna. The pronghorn antelope is able to run at over 60 km per hour (40 mph), which is much faster than any modern American predator, but it may have developed as a strategy to outrun a large running cat like the American cheetah, in much the same way as Thomson's gazelles flee modern cheetahs today. There are also trees in Central America that produce huge quantities of fruit each year, so much that the local dispersers such as peccaries and agoutis cannot possibly eat them all. Is this an example of the trees being inefficient (after all, why put all that energy into something that gives you little or no return?), or did they produce this fruit so that gomphotheres, ground sloths and other megafauna could consume them and then excrete them in their dung some distance away from the parent tree? Comparison with the shape of fruits eaten by elephants today in tropical Africa suggests that the latter is more likely.

If all these animals still survived in the Americas and Australia they would surely attract filmmakers and tourists from all over the world, in much the same way that people go on safari in Africa today. No matter what the causes of their extinction were, it is interesting to wonder what the world would be like today if all these animals still existed. Considering the problems we have trying to preserve the modern megafauna, imagine the burden on our consciences if there were eight species of elephant to save rather than two.

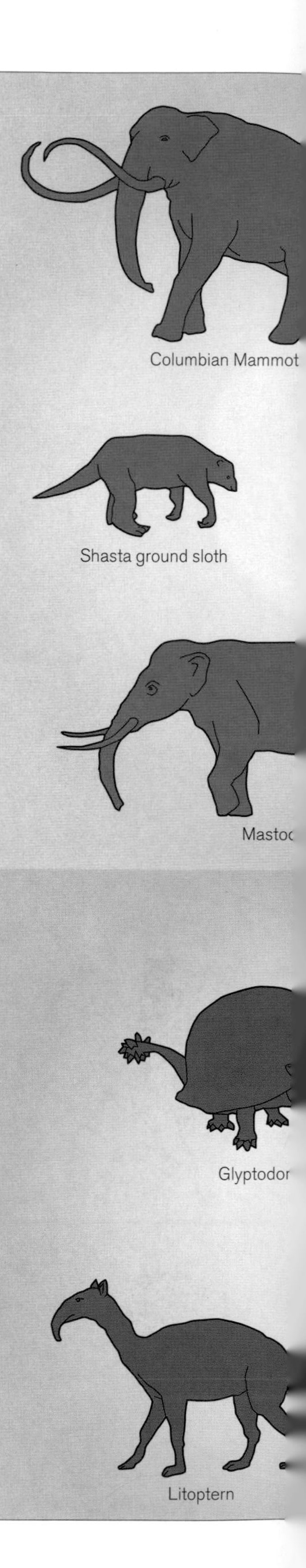

The scale of large mammal extinction towards the end of the Ice Age was enormous. Animals on all continents were affected, although those in Africa and Europe suffered less than those in the Americas and Australia. Not all animals were equally affected, rodents and other small creatures tended to survive, while the numbers of large mammals lost is quite staggering – 50 species in South America alone.

Yesterday's Camel
Cave bear
Woolly Mammoth
Giant ground sloth
Sabre-toothed cat
Woolly rhinoceros
Giant deer
Marsupial lion
Giant kangaroo
Notoungulate
Diprotodon
Short-faced kangaroo
Giant short-faced kangaroo

7

After the Ice

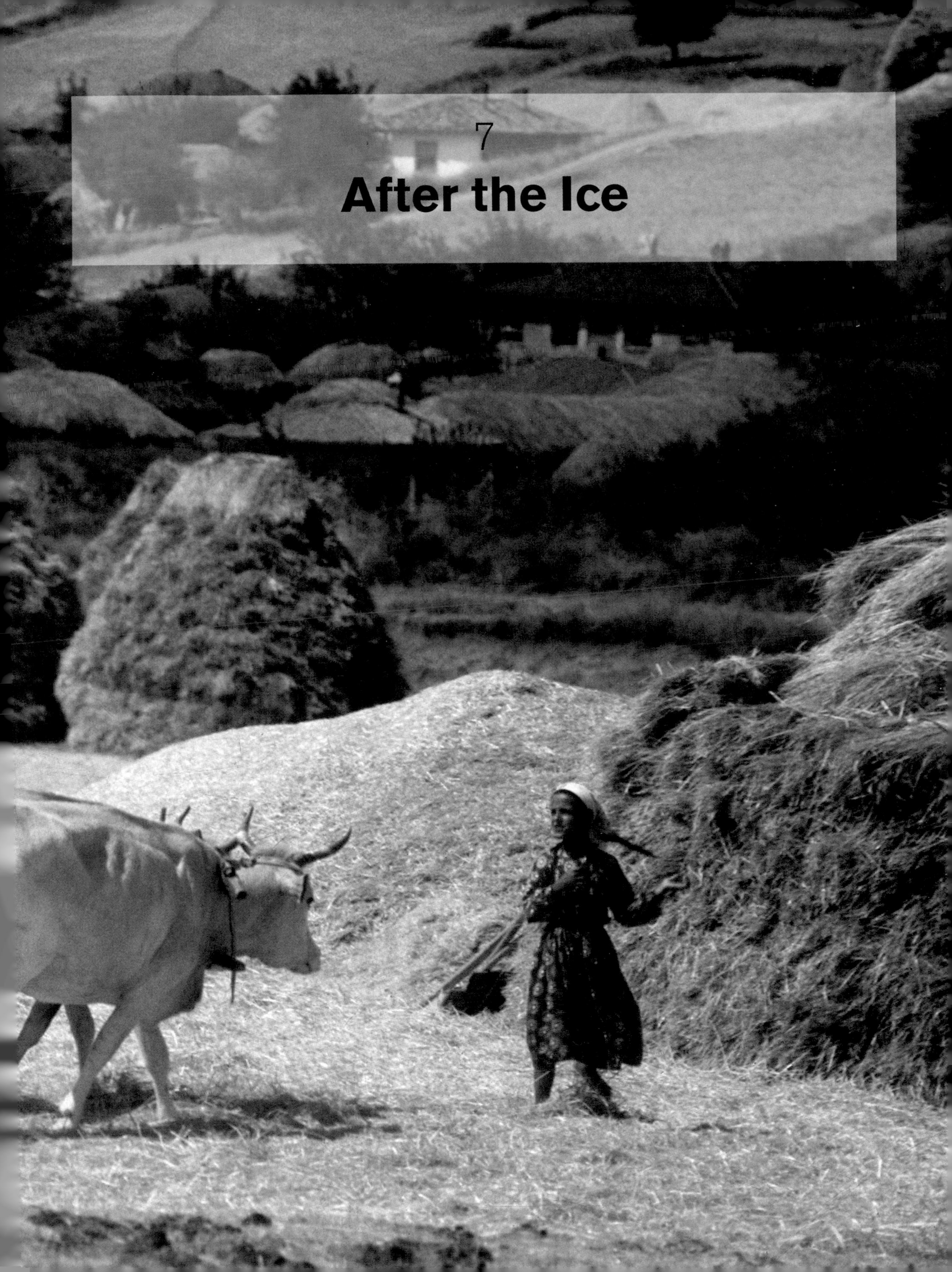

GEOLOGISTS REFER TO THE TEN OR SO MILLENNIA since the last glacial period of the Ice Age as the 'Holocene' (from the Greek *holos*, meaning 'entire' and *kainos*, 'recent', the word thus meaning 'entirely recent'). The subdivision is completely artificial, for we are, in fact, in an interglacial, albeit one that, so far, shows no signs of ending in another glacial episode. In the natural order of things, this should occur in the future, although quite when is anyone's guess, and indeed humanly caused warming may irreparably upset the natural glacial–interglacial cycle (see Chapter 8).

previous pages Food production was the seminal innovation after the Ice Age. Millions of farmers, such as this Turkish family, still practise subsistence agriculture today, living from harvest to harvest.

When the last major cold episode of the Ice Age ended about 15,000 years ago, widespread global warming ensued, followed by an abrupt return to near-glacial conditions around 10,950 BC. This 1,100-year-long cold snap, known to geologists as the 'Younger Dryas' (named after an alpine flower) ended as abruptly as it began with the renewed onset of warmer conditions that persisted until about 5,000 years ago, since when there has been a slightly cooler and drier global climate, resulting in a massive expansion of the Sahara desert. It seems that only fossil fuels being pumped into the atmosphere during the Industrial Revolution has reversed this cooling trend. Although the climate since the end of the Younger Dryas may appear relatively stable, throughout the past 10,000 years there have been constant minor oscillations. Known as 'Bond events', these oscillations are the equivalent of similar 'Dansgaard–Oeschger cycles' during the Ice Age (see Chapter 4), and seem to occur every 1,500 years or so. They are brief cold periods, which seem to be connected to the rapid 'switching off' of the formation of deep ocean water in the Greenland Sea and have been linked to droughts in the Middle East and disruption of the Southeast Asian monsoons. Controversially, the Bond event at 2200 BC has been said to have significantly affected ancient societies in a number of Eurasian countries, including Egypt, Mesopotamia, Anatolia, Greece, Israel, India, Afghanistan and China. The last Bond event was the period often referred to as the Little Ice Age, which had a profound effect on the people of northern Europe (p.204).

During the centuries of dramatic warming that occurred after the last glacial, hunter-gatherer societies throughout the world had to contend with staggering environmental changes, which are unimaginable by modern standards. These included sea-level rises in the order of 90–120 m (300–400 ft) (compare this with the 0.3- to 0.8-m (1- to 2.6-ft) rise predicted for the next century), the rapid spread of forests throughout much of Europe and parts of Eurasia, the severing of land bridges such as the one joining Alaska and Siberia, and rapid shrinking of ice sheets in North America, Scandinavia and the Alps. Human societies everywhere adapted to these fundamental environmental shifts with remarkable flexibility. They changed their ways of life by intensifying and specializing the food quest, before adopting a completely new form of subsistence – agriculture and animal domestication. For the first time in human existence, people began to produce their own food. For the past 10,000 years our survival has depended on farming, and thus on higher food outputs that have fed growing populations and increasingly elaborate civilizations. Today, in one of the paradoxes of history, our reckless pursuit of growth and more lavish lifestyles is triggering a global warming that threatens the agriculture upon which our survival depends.

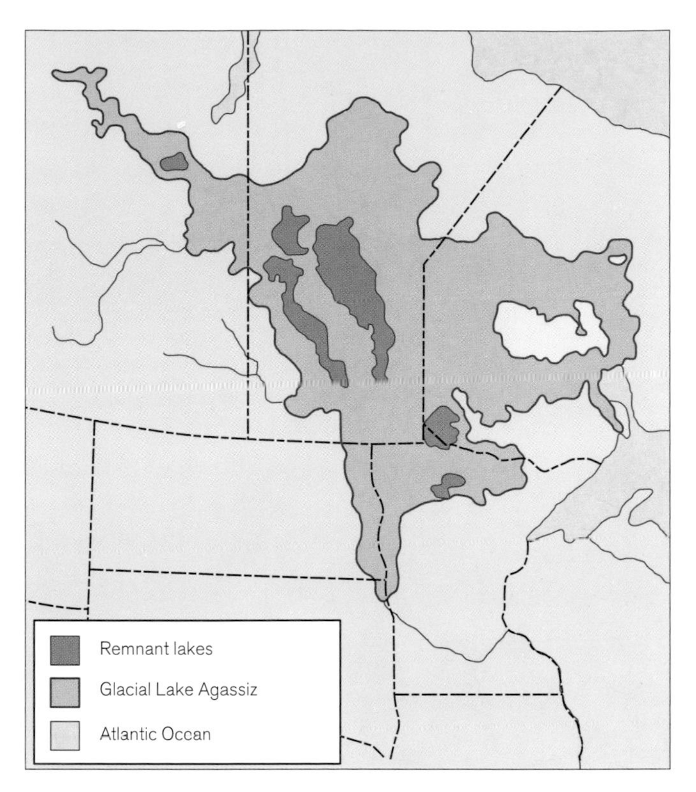

Glacial Lake Agassiz formed over much of what is now central Canada and the upper Midwest of the United States as the vast Laurentide ice sheet melted rapidly at the end of the last glacial period. In 12,000 BC, its heaving waters lapped the retreating ice sheet for 1,100 km (683 miles). A southward bulge of the ice formed a peninsula that blocked Lake Agassiz's waters from flowing down what is now the St Lawrence Valley.

The last cold snap

The cold period known as the Younger Dryas came about when Lake Agassiz, a huge glacial lake in Manitoba and the upper Midwest, which had been formed by melting ice from the retreating North American ice sheets, broke its banks. Billions of litres of freshwater cascaded into the North Atlantic Ocean and flowed across the surface of the denser salt water. This abruptly slowed, or may even have stopped, the natural downwelling of the warmer waters of the Gulf Stream off southern Greenland. Within a generation or so, Europe was plunged into near-arctic conditions. North America suffered the same fate. Severe drought gripped much of southwestern Asia. After eleven centuries, the Atlantic and the Gulf Stream absorbed the effects of the freshwater and the Gulf Stream gained full strength once more.

Widespread global warming transformed global geography before and after the Younger Dryas. The ice sheets that had mantled Scandinavia and the Alps retreated, Canada gradually emerged from its ice cover and sea levels rose irregularly but rapidly throughout the world. The sea level at the height of the late Ice Age was about 120 m (394 ft) below modern levels. By 10,500 BC, the world's oceans had risen by 20 m (66 ft). There was then an abrupt rise of 24 m (79 ft) in a mere 1,000 years, another slight rise at about 9000 BC, and a 28-m (92-ft) rise between 8500 and 7500 BC.

Retreating ice sheets opened up new areas for human settlement in Scandinavia and North America, while rising sea levels severed the Bering Land Bridge between Siberia and Alaska by as early as 9000 BC. The flat lands of what is now the southern North Sea vanished. Britain separated from the continent by about 6000 BC. The great continental shelf off Southeast Asia vanished under the rising Pacific, as did coastal plains off long stretches of the North American coast.

Throughout the world, late Ice Age hunter-gatherer populations settled in newly exposed lands or moved to higher ground. In places like Tasmania, people now lived in complete isolation until modern times. The millennia of warming were a time of major adjustment for human societies in many parts of the world.

Eurasia

The warming at the end of the Ice Age was rapid and irregular, but in many ways no different from that of earlier times. For thousands of years, people had scheduled their lives around reindeer migrations and salmon runs in spring and autumn, and had taken advantage of the short growing seasons of plant foods. Now the rhythms of people's lives adjusted to new realities. The reindeer herds moved northwards with the now-shrinking tundra as temperatures increased. Some groups followed them to open plains south of what was to become the Baltic Sea. Theirs was a mobile way of life, anchored to glacial lakes and river valleys. Small bands ambushed reindeer as they migrated in spring and autumn, using bows and arrows to kill dozens of animals near lakes and in narrow valleys.

Other groups of people stayed where they were, in what soon became forested landscapes in France, Germany and Spain. The centuries of warming brought dramatic environmental changes to ancient hunting grounds; pine and birch trees

Hundreds of caribou migrate northwards in northern Canada. For thousands of years during the Ice Age, vast caribou and reindeer herds migrated north and south to and from the tundra. The autumn migration, when the beasts were fat and in good condition, was the best time for both Ice Age and later hunters to harvest migrating animals for antler, fat, hide, meat and other by-products.

expanded into valleys and onto plains, then oak and other trees gradually formed the primordial forests that were to mantle Europe until they were finally decimated by Roman and medieval farmers. The familiar mammoth, woolly rhinoceros and steppe bison of the late Ice Age soon disappeared in the face of warmer temperatures. The hunters now focused on small and medium-sized forest animals, such as the red deer, the rare giant stag and the wild boar. At the same time, birds, fish and plant foods assumed greater importance in people's diet.

Mesolithic flint microliths from Kelling Heath, Norfolk, England. These tiny points were attached to spears and arrows to form lethal barbs for killing prey.

The changeover to small game did not require any drastic alterations in hunting technology, except, in many areas, for a shift from antler and bone to wood for most tools and weapons. The Magdalenians of western Europe and their contemporaries to the east had developed increasingly sophisticated, lighter weaponry in earlier times. They may even have relied on the bow and arrow, although the evidence is uncertain. Lightweight hunting weapons armed with razor-sharp stone barbs were already in use, and became increasingly important as temperatures warmed up and the emphasis changed from herd animals like reindeer and saiga to individual prey, often hunted at the edge of clearings or in the forest. Hunters now relied on traps and snares, on arrows loosed at solitary animals stalked cautiously from tree to tree. Birds of all kinds became an important food source, waterfowl taken with nets and snares, others shot on the wing with light arrows armed with diminutive stone points.

These were the millennia when the bow and arrow came into its own. The vagaries of preservation are such that we have virtually no evidence except the tiny flint 'microliths' used as lethal points and barbs on spears and arrows. The hunters fabricated them by the hundred, shaping tiny points and barbs on small flint blades struck from carefully shaped nodules, then notching them and snapping off the thicker base before mounting them in wooden points. Microliths alone hardly give a comprehensive impression of human societies. (The term 'Mesolithic' – Middle Stone Age – from the Greek *mesos*, 'middle', and *lithos*, 'stone', is sometimes used to describe these societies.) The preserved weapons give us a misleading impression of

above The Iron Gate of the Danube, where the river passes through the Carpathian Mountains. Hunter-gatherer groups preyed on fish and game here for thousands of years, eventually camping in permanent settlements like Lepenski Vir. This important site dates to about 6500 BC.

seemingly culturally impoverished human societies that now relied heavily on wood for artifacts of all kinds. The archaeologists of three quarters of a century ago spoke of post-Ice Age times as a 'hiatus', as it were, before agriculture and other innovations transformed European history. Nothing could be further from the truth. In fact, Europeans everywhere adapted skilfully to rapidly changing environments and new food resources that were scattered in patches over very diverse, warming landscapes. Above all, they settled in forest clearings, by lake shores and river banks, and along newly exposed sea coasts.

During the late Ice Age, the densest human populations had flourished in sheltered locales, where deep valleys and a wide variety of game and other foods were relatively abundant. France's Vezère Valley and the Danube Valley in central Europe are two examples. Now the densest populations were to be found along rivers, at strategic locations like the Danube's Iron Gates, and along freshwater lakes where fish such as pike abounded.

Some of the greatest changes resulted from fast-rising sea levels, which flooded continental shelves, inundated estuaries and turned many once rapid-flowing rivers into sluggish streams. The sea-level rise created rich shoreline environments, abundant in all kinds of fish and shellfish, in sheltered estuaries along the Atlantic coast, in the North Sea region, and along the coasts of the Baltic Sea. The Baltic developed originally as a glacial lake, formed by retreating Scandinavian ice sheets, eventually becoming a brackish sea joined to the Atlantic by the Skagerrak. As early as 8000 BC, hunter-gatherer bands settled along Baltic shores, collecting shellfish, spearing pike and other fish with barbed antler, bone and wooden spears, and harvesting catches with fibre nets and traps set in shallow water. These 'Maglemose' and, later, 'Ertebølle', societies exploited increasingly smaller territories, spent more time in more permanent settlements and developed more elaborate social institutions, reflecting more crowded landscapes and increasing competition for foods of all kinds. There is increasing evidence for inter-group fighting too, witness the spear points found embedded in bodies buried in local cemeteries.

left A sculpted head from Lepenski Vir, thought to depict a mythic being, perhaps a fish god. These sculptures lay in house foundations.

As was the case along Baltic shores, people now often relied on intensive exploitation of locally abundant and predictable resources such as salmon or nuts. Seasonal phenomena such as salmon runs, game migrations or hickory nut harvests required not only the harvesting of enormous quantities of food in a short time but also their processing and storage for later use. Storage technology now assumed a new importance; thousands of fish were dried on racks in the sun or in front of fires, and nuts and wild cereal harvests were placed in basket- or clay-lined pits for later consumption. There was nothing new in the notion of storage; much earlier big-game hunters, for example, dried meat and pounded it up to eat on the march. What was new, however, was large-scale storage in more sedentary settlements, where mobility was no longer a viable strategy. By the use of storage and by the careful seasonal 'mapping' of game, plant and aquatic resources, post-Ice Age hunter-gatherers compensated for periodic food shortages caused by short-term climatic change and seasonal fluctuations. At the same time, they broadened their diet to include foods that were less nourishing, another way of ensuring fallback in the diet in the event of staple shortages.

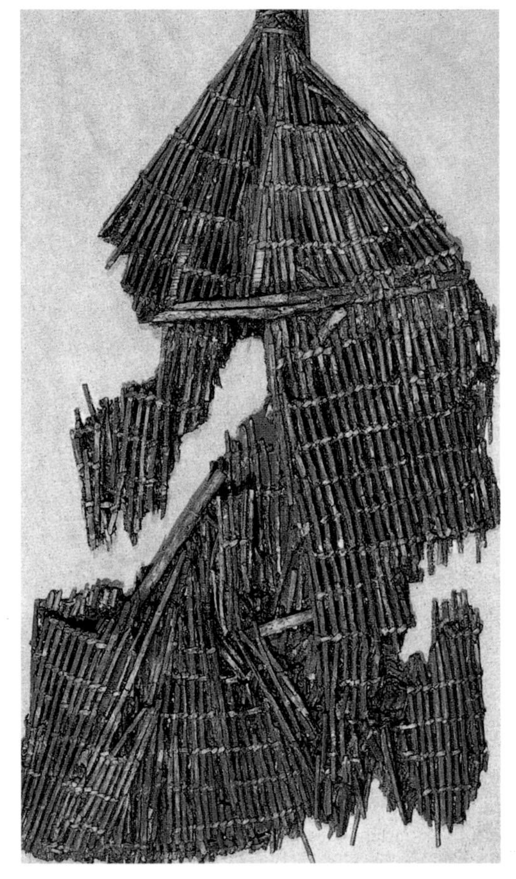

A wooden fish trap from the Little Knabstrup site in Denmark. The slats of the trap are bound together with reeds. Its maker belonged to the Ertebølle culture, late 5th millennium BC.

Southwestern Asia

During the late Ice Age, small bands of hunter-gatherers ranged over large hunting territories from the eastern Mediterranean coast far inland. These people, known to archaeologists as 'Kebarans', exploited both forested and semi-arid environments. They may have moved to the uplands during the hot summer months, relying heavily on such animals as the desert gazelle for much of their diet.

By about 13,000 BC, warming temperatures throughout southwestern Asia brought significant environmental and vegetational changes. In the late Ice Age, warmth-loving plants like wild emmer wheat and barley, oaks, almonds, and pistachio trees were restricted to refuge areas lying below the modern sea level. These sandy-soil areas would have given only poor yields of wild cereals. As early Holocene temperatures rose, these plant and tree species colonized higher country, where clayey soils produced much higher yields. New, denser wild cereal stands were now much more resistant to short-term climatic change and were also harvestable over a longer period of time each year. Many groups now shifted to consuming mainly plant foods. We know this because their settlements are littered with ground stone tools – pestles and mortars, also implements used to process seed harvests for storage, essential in much more seasonal, and increasingly arid, climates.

above Ground stone pestles from the Natufian culture of the Near East, c. 11,000 BC. Grinding stones and mortars soon became a fundamental part of early farming societies. Women spent much of their day laboriously grinding grain with such artifacts.

These abundant cereal and nut resources within the Mediterranean hill zone stimulated the development of more intensive foraging strategies, more sedentary settlement and rapid territorial expansion until the most favoured territories were filled. New societies, known generically as the 'Natufian' culture, emerged from Kebaran roots in hilly areas. Here wild cereals and nut-bearing trees had their natural habitat. The larger sites are close to the boundaries between the coastal plains or grassland valleys and the hill zone, and some were placed strategically to take advantage of good toolmaking stone. Such settlements enabled the Natufians to exploit spring cereal crops, autumn nut harvests and the game that flourished on the lowlands and on the rich nut mast on the forest floor in the hills above. Unlike their more mobile ancestors, the Natufians enjoyed many months of plentiful food by exploiting spring cereals and then following nut harvests up the slopes as the nuts ripened at progressively higher elevations. Gazelle hunting assumed great importance at certain seasons of the year, with neighbouring communities cooperating in game drives, ambushes and other mass-hunting enterprises.

By the end of Natufian times, about 10,500 BC, local populations were considerably higher than in earlier times. Natufian society offers intriguing glimpses of a new, more complex social order. The Natufians buried their dead in cemeteries,

opposite Natufian burial from the Wadi Me'arot site in Israel. The cave contains Neanderthal and other early occupations as well as Natufian levels dating to before 10,000 years ago.

which have yielded a wealth of information about their society. There are clear signs of social ranking. One common and constantly recurring symbolic artifact, the *dentalium* seashell, is confined to a few burials, whereas the elaborate grave furniture, such as the stone bowls found with some individuals, including children, hints strongly at some form of social differences. Perhaps this social ranking was the result of a need for the redistribution of food surpluses and to maintain order within much larger sedentary communities. Also, the stone slab grave covers and mortar markers associated with the cemeteries may have served as ritual markers of territorial boundaries, perhaps of lands vested in revered ancestors.

The 1,100 years of the Younger Dryas brought savage droughts to southwestern Asia, while Europe was plunged into near-glacial conditions once again. After 11,000 BC, the Natufians faced much drier climatic conditions at a time when their populations were expanding. Increased aridity shrank the cereal habitats in the Mediterranean zone, causing the most productive stands to be found at higher altitudes. At the same time, the Natufians were forced to remain in sedentary settlements close to permanent water supplies, a fact of life that made the cost of harvesting cereals and nuts at remote locations much higher. It was these conditions that helped give rise to what was to prove a momentous development – the initiation of farming.

A bone sickle handle used by a Natufian to gather wild grasses, from the Kebara site on the western side of Mt Carmel, Israel, c. 10,000 BC. A row of sharp flint blades set in the handle provided an effective and easily sharpened tool.

The first farmers

How could people living in a complex hunter-gatherer society solve the problem of declining staples? After nearly 2,000 years of close involvement with cereal plants, people would have been well aware of what was needed to plant and grow cereal grasses deliberately. By deliberately planting cereals on a modest scale, the people tried to cope with uncertainty, to augment declining stands of wild wheat and barley with their own supplementary crops.

This was the moment when agriculture and animal domestication entered human history. People began to harvest wild cereal grasses, such as rye, with sickles or by uprooting entire plants. The harvesters imposed entirely new selective pressures on cultivated stands of wild grasses. This new set of selective pressures seems to have developed first in a small region of southwest Asia called the Levantine Corridor, an area up to 40 km (25 miles) wide running from the Damascus Basin in the North into the lower Jericho Valley, and extending into the Euphrates Valley. This corridor had reliable water supplies and a relatively high water table, enabling foragers to shift wild grains from their natural habitats into well-watered areas near streams and lakes. The Abu Hureyra mound in the Euphrates Valley chronicles a dramatic shift from hunting and gathering to agriculture at around 10,000 BC, when a sedentary village of semi-subterranean dwellings suddenly became a farming village of mudbrick houses anchored to fields of einkorn and rye. A thousand years later, the inhabitants maintained herds of sheep and goats. As the Younger Dryas ended, the new economies spread widely through southwestern Asia like wildfire.

left A plastered skull from Jericho in Jordan. Such artifacts were probably associated with an ancestor cult.

These were times of profound change in human society – permanent settlement in villages, and soon larger villages or even small towns at places like Jericho in Jordan, with its famous walled settlement, and Çatalhöyük in central Turkey, where elaborate shrines commemorated the ancestors. Agriculture ushered in new relationships with the land, where the ancestors served as a crucial spiritual link between the living and the powers of the supernatural world. At the same time, kin ties and relationships with neighbouring communities assumed greater importance, as trade in essential commodities expanded dramatically. Farming populations rose fast, especially in areas of exceptional fertility such as the Nile Valley and southern Mesopotamia.

below Rice cultivation in the Yangtze Valley in China, a centre of early rice domestication long before 6000 BC.

The droughts of the Younger Dryas played a significant role in the appearance of agriculture, which probably began not as a revolutionary invention, but as a logical response to aridity, aimed at preserving a familiar way of life. In the event, the genetic changes in cereals, and later in animals, were so rapid and successful that the new economies replaced hunting and gathering over enormous areas of southwestern Asia. By 6000 BC, farming was spreading into Europe and cattle were being herded in the then somewhat wetter Sahara.

Agriculture began independently in many places at different times – in the Indus Valley region of India by at least 6000 BC, in northern China at about the same time, and in China's Yangtze Valley with rice cultivation by at least 7000 BC. In all these areas, people were well aware that wild grasses germinated and they had harvested them for thousands of years. When drier conditions took hold and populations increased, people began cultivating the same cereals they had foraged and soon became full-time farmers.

The Americas

As the world climate warmed up, the American landscape changed drastically. If one word can be used to describe these profound changes, it is 'diversity'. A great variety of local environments emerged: lush river floodplains, great deserts, grassy plains and miles upon miles of boreal and deciduous woodland.

In general, the western and southwestern United States became drier, but the East Coast and much of the Midwest became densely forested. Most of the large Pleistocene mammals became extinct, but the bison remained a major source of food, especially on the High Plains. In the West, the warmer climate brought drier conditions, which meant less standing water, markedly seasonal rainfall, and specialization among humans in fishing or intensive gathering, where possible.

above A projectile point embedded in a bison skull from a Paleo-Indian kill site at Farson, Wyoming.

The human population of North America was still sparse, scattered in isolated hunter-gatherer bands. Judging from sites in many areas, people spent most of the year living in small family groups and exploiting large hunting territories. They may have come together with their neighbours for a few weeks during the summer months at favoured locations near rivers or nut groves. At first there would probably have been plenty of vacant territory to go around. In time, natural population growth and the low carrying capacity of the land combined to restrict mobility. Just as in the Old World, hunters now focused on smaller mammals, especially the white-tailed deer. Inevitably, too, people turned to alternative food sources, including plants, birds, molluscs and fish. From the earliest times, Native Americans developed expert knowledge of wild plant foods, which preadapted them in later times to the cultivation of native grasses and tubers. After 2000 BC, many groups, especially those in midwestern river valleys, now lived in base camps occupied for many months of the year. These camps served as anchors for larger territories, which were exploited seasonally from outlying settlements.

Such was the diversity of North American environments in Holocene times that only a few relatively limited areas witnessed sedentary settlement and large-scale population growth. Perhaps it is no coincidence that it was in some of these areas, notably midwestern and southeastern river valleys and the southwest, that people turned to the deliberate cultivation of native plants.

In Central and South America, there was a similar intensification of hunting and gathering, especially in the exploitation of wild plant foods of all kinds. This intensification was in response both to the extinction of larger animals after the Ice Age and the new warmer and drier climatic conditions. Grasses and edible roots of all kinds assumed increasingly greater importance in the human diet, not only in tropical forests like those of Panama and the Amazon Basin, but also in drier, more open environments and high in the Andes.

Native Americans domesticated an impressive range of native plants, including maize (domesticated from a native grass, teosinte, perhaps in Central America's tropical lowlands), beans and root crops like cassava and the potato. Animal domestication occurred somewhat later, partly because there were very few potential domesticates, unlike in the Old World, where there were 26. Plant domestication began among hunter-gatherer societies in tropical and semi-tropical regions of Central and South America, probably as a result of selective pressures on wild plants caused by intensive harvesting. The first plants were domesticated by at least 5000 BC, if not earlier. Well before that date native Central and South Americans had developed a remarkable expertise with plants of all kinds, to the point that it was almost inevitable that they would begin cultivating them, both to increase food supplies and as a way of preserving their traditional ways of life in the face of changing environmental conditions. Squashes were probably the first plants to be domesticated, followed by potatoes in the Andean highlands, then maize and beans.

above Maize (*Zea mays*) was the staple of Pre-Columbian farmers, cultivated from arid South America, the Amazon Basin and the Andes northwards through Central America right up to the St Lawrence River.

left The Elizabethan artist John White depicted two Algonquin Indians in Virginia at their evening meal in the late 16th century AD. A fellow colonist, John Heriot, wrote of the Indians: 'They are verye sober in their eating and trinkinge, and consequently verye longe lived because they doe not oppress nature . . . I would to God we would followe their exemple.'

Just as it had in the Old World, farming spread rapidly throughout the New World, to become near-universal in many environments within about 2,000 years.

When Europeans arrived, maize was the staff of life for hundreds of native American societies, from Pueblo societies in the American southwest to Iroquois villages along the St Lawrence River, and in every kind of warmer environment imaginable throughout Central and South America.

above John White's depiction of an Algonquin village shows houses, a ceremonial structure, and maize fields growing right up to the dwellings. White portrayed an idealized view of Native American life, which did not reflect the realities of a factionalized, often violent society.

The emergence of civilization

In about 15,000 BC, the Persian Gulf was dry land, so the Tigris and Euphrates rivers flowed into the Gulf of Oman 800 km (500 miles) south of their present mouths. After 12,000 BC, sea levels rose rapidly, as they did in other parts of the world, from about 20 m (66 ft) below modern levels in 6000 BC to 2 m (7 ft) above modern levels between 4000 and 3000 BC. As sea levels rose, riverborne alluvium filled the Persian Gulf, building a vast delta that advanced gradually to the south, leaving freshwater lakes and marshes in its wake. Both hunter-gatherers and farmers lived off fish, game and plant foods in these fertile, low-lying areas for many thousands of years.

Before 6000 BC, villages in southern Mesopotamia supported populations of between 50 and 200 people. In about 6000 BC, a population increase began; people now lived in much larger towns, and much of the landscape was otherwise devoid of settlement. At the same time, contacts between neighbouring communities

intensified, perhaps in part to make up for shortages of grain and other commodities in different areas during a time of climatic change. Within a millennium, there were large cities in Mesopotamia. These changes took place as the Mesopotamian landscape altered dramatically. Rising sea levels forced settlements to shift frequently, rainfall became more irregular, the environment became drier, and agricultural populations were dislocated.

above An aerial shot of the central precincts of Uruk in Mesopotamia, a city before 4000 BC, and later a major Sumerian commercial and religious centre.

By the mid-4th millennium BC, the climate and land had stabilized, but there was now less rainfall, so that farmers in southern Mesopotamia depended for much of their water on the annual floods of the Euphrates and Tigris. Carefully organized irrigation agriculture now became essential and developed rapidly from small, informal canal systems that diverted flood water onto otherwise arid, but fertile soils. Such agriculture could feed much larger populations, but depended on intensive labour. Villagers now abandoned dispersed, small communities and moved to favoured areas. They developed large-scale irrigation systems and contributed to a new economic infrastructure. Within a short time, a new demographic and social order developed that culminated in the appearance of the world's first literate urban civilization, that of the Sumerians, in about 3100 BC.

Cities and civilizations soon appeared elsewhere – in Egypt by 3100 BC, in the Indus Valley by 2600 BC, and in northern China by about the same time. All of these pre-industrial states displayed the same general characteristics – strongly centralized, hierarchical governments usually associated with densely populated cities, and social stratification, where thousands laboured for the benefit of a few, the pinnacle of society. The same long-term trend towards increasing social complexity

below The central precincts of the Maya city of Tikal, which presided over a powerful domain for much of the first millennium AD. Maya civilization flourished in the Mesoamerican lowlands from the 1st millennium BC and lasted until the Spanish Conquest in the early 16th century AD. During the 10th century, Maya civilization in the southern lowlands collapsed. Great cities like Tikal and Copán imploded, partly as a result of drought and an overstressed environment.

developed in the Americas – as early as 3000 BC in coastal Peru, and over widespread areas of Central and South America within the next thousand years, culminating in the elaborate Maya, Toltec and Aztec civilizations of Central America and the Moche, Tiwanaku and Inca of the Andes.

Our ancestors of more recent times experienced minor but significant shifts in climate, among them the warmer centuries of the Roman era, which allowed the Romans to turn much of western Europe into a huge granary to feed their armies and cities. The so-called 'Medieval Warm Period' of about AD 800 to 1250, which is still ill-defined, brought higher summer temperatures, long dry spells and many years of bountiful harvests to western Europe. Medieval populations rose, marginal lands were taken into production, and thousands of hectares of forests cleared at a time when people still lived at the subsistence level, from one harvest to the next. Vineyards flourished in central England; cereal crops grew in Norway. Favourable ice conditions in the north made it easier for Norse voyagers to colonize Iceland and Greenland, and to cross regularly to Labrador. The warmer centuries were less kind to the Americas. Prolonged droughts settled over western North America; Maya civilization partially collapsed under a series of severe arid spells; dry La Niña-like

below Pueblo Bonito in the Southwest USA, the greatest of Chaco Canyon's pueblos, or 'great houses', built by the Ancestral Pueblo between the 9th and 12th centuries AD. This multi-storey pueblo with its enclosed plazas and subterranean ceremonial chambers or *kivas* (the circular structures in the picture) was a major ceremonial centre for many generations. The Ancestral Pueblo abandoned Chaco Canyon partly as a result of long-term drought in the 12th century.

conditions lingered over much of the eastern Pacific. There were droughts in Mongolia and eastern Africa.

During the 14th century, climatic conditions became cooler and more unsettled, with increased storminess in Europe, with the onset of the period known as the 'Little Ice Age'. A seven-year period of heavy summer rains and poor harvests killed over 1.5 million subsistence farmers. Storm surges and severe gales in the North Sea killed thousands of people in the Low Countries during the 14th and 15th centuries. The climax of the Little Ice Age came in the late 17th century, during the Maunder Minimum, a period of reduced sunspot activity. This was when the Thames froze over and fairs thrived on the ice; these were to become the signature events of the Little Ice Age. The cooler and unsettled centuries saw major fluctuations in Alpine glaciers, as well as a revolution in agriculture that first took hold in the Netherlands, then Britain. The incidence of famine was reduced, except in much of France, where bread shortages, caused in part by poor harvests, contributed to the unrest of the French Revolution.

In the mid-19th century, the world entered a new era – one in which humanly caused global warming is changing climatic equations. As Mark Maslin points out in Chapter 8, average global temperatures have risen over 0.75°C (1.3°F) over the past 150 years and sea levels have risen over 22 cm (8.6 inches), in large part because of an increase in greenhouse gases in the atmosphere. We are entering a future where the impacts of global warming will intensify, with significant, often threatening rapidity.

Until recently, the Holocene has been a time of relatively stable climatic conditions, but we know that even minor temperature and rainfall shifts had dramatic effects on human societies. Such changes played a role in the development of agriculture and animal domestication, and in the appearance of the world's first civilizations. Monsoon failures triggered by El Niño events during the 19th century remind us that even short-term climatic events can wreak havoc with human societies. One estimate calculates that between 20 and 30 million tropical farmers perished from famine and famine-related diseases during that century as a result. Throughout history, human societies, however complex, have been at the mercy of climatic change. We would be rash to assume that our own huge cities are impregnable to climatic forces that are beyond our control.

A frost fair on the Thames, commemorated by the Dutch artist Abraham-Danielz Hondius (1625–1695). The Little Ice Age is epitomized by the major freezes that caused the Thames to ice over, especially during the late 17th century. Impromptu frost fairs flourished on the frozen river, complete with dancing bear baiting and other amusements. Thames watermen suffered greatly from such freezes and were instrumental in organizing the fairs, which were very profitable.

8
Hot or Cold Future?

We now know that for the last 2.5 million years the great Ice Age has waxed and waned. Global climate has cycled from conditions that were similar or even slightly warmer than today, to full glaciations, which caused ice sheets over 3 km (1.8 miles) thick to form over much of North America and Europe. As we have seen, these Ice Age cycles are driven primarily by changes in the earth's orbit with respect to the sun. In fact the world has spent over 80 per cent of the last 2.5 million years in conditions much colder than the present. We are now in a warm period or interglacial, and sometime in the future the world should plunge into the next glacial period. Nowadays, however, you cannot open a newspaper without being hit with headlines about global warming. This chapter shows why we now face human-induced global warming rather than an impending glaciation. Ironically, the fear of the next ice age did mean that the threat of global warming was not recognized until the late 1980s. In this chapter we will also investigate why the 'ice' legacy of previous glaciations has made earth's climate so sensitive to the additional carbon dioxide added through burning of fossil fuels.

Every second we burn 170 million kg (375 million lb) of coal, 1,252 million litres (330 million gallons) of gas and 110,692 litres (29,000 gallons) of oil, which combined with deforestation results in 1.8 billion tons of carbon dioxide released every year into our atmosphere.

Why the delay in recognizing global warming?

How did the fear of the next glacial stop us from recognizing global warming? Spencer Weart, director of the Center of History of Physics at the American Institute of Physics, argues that all the scientific facts about increased atmospheric carbon dioxide and potential global warming were assembled by the late 1950s to early 1960s. He argues that the political Cold War environment favoured the funding of physical geosciences which led to so much of the fundamental work on global warming being completed. Gilbert Plass published an article in 1959 in *Scientific American* declaring that the world's temperature would rise by 3°C (5.4°F) by the end of the century. The magazine editors published an accompanying photograph of coal smoke belching from factories and the caption read: 'Man upsets the balance of natural processes by adding billions of tons of carbon dioxide to the atmosphere each year'. This resembles thousands of magazine articles, television news items and documentaries that we have all seen since the late 1980s. This delay between the science of global warming being accepted and in place in the late 1950s and early 1960s, and the sudden realization of the true threat of global warming during the late 1980s, was in part due to the fear of the supposed impending glacial period.

The worry about the next glaciation was a result of the power of what is known as the global mean temperature data set. This is calculated using the land–air and sea–surface temperature. From 1940 until the mid-1970s the global temperature curve seems to have had a general downward trend (see figure below). This provoked many scientists to discuss whether the earth was entering the next glacial period. This fear developed in part because of increased awareness in the 1970s of how variable global climate had been in the past. The emerging subject of palaeoceanography (study of past oceans) demonstrated from deep-sea sediments that there were at least 32 glacial–interglacial (cold–warm) cycles in the last 2.5 million years, not four as had been previously assumed. The time resolution of these studies was low, so that there was no possibility of estimating how quickly the glacials came and went, only how regularly. It led many scientists and the media to ignore the scientific revelations of the 1950s and 1960s and to talk instead of global cooling. As Lowell Ponte (1976) summarized:

> Since the 1940s the northern half of our planet has been cooling rapidly. Already the effect in the United States is the same as if every city had been picked up by giant hands and set down more than 100 miles closer to the North Pole. If the cooling continues, warned the National Academy of Sciences in 1975, we could possibly witness the

previous pages Future climate change threatens some of the world's most magnificent habitats. Most at risk are the Arctic and Antarctic, as the warming there has already started to melt the ice. Thousands of species have already become extinct and many more are close to it, including iconic animals such as the polar bear.

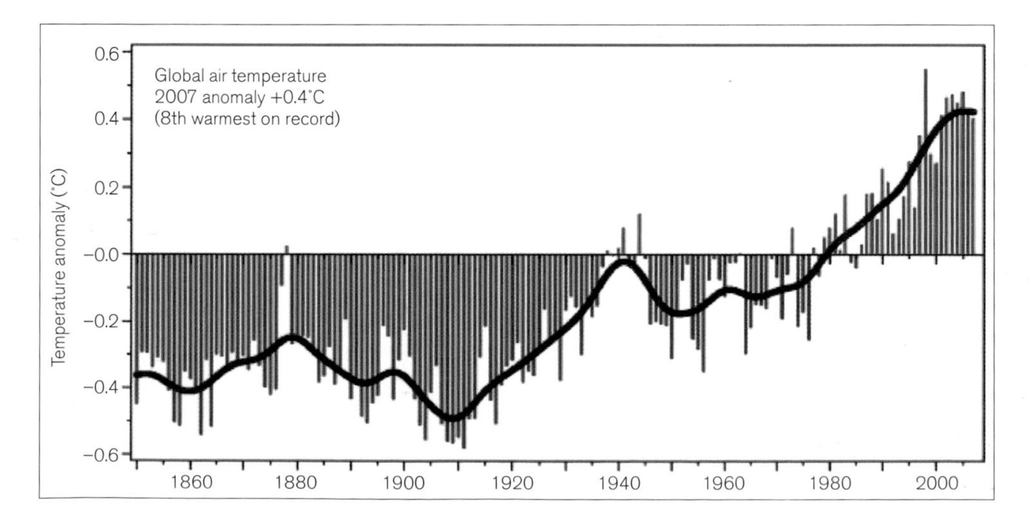

left Over the past 150 years the earth has warmed by over 0.75°C (1.3°F). The 12 warmest years on record have all occurred in the last 13 years, with 2005 the warmest, and 1998, 2002, 2003 and 2004 second, third, fourth and fifth warmest respectively.

beginning of the next great Ice Age. Conceivably, some of us might live to see huge snow fields remaining year-round in northern regions of the United States and Europe. Probably, we would see mass global famine in our lifetimes, perhaps even within a decade. Since 1970, half a million human beings in northern Africa and Asia have starved because of floods and droughts caused by the cooling climate.

It was not until the early 1980s, when the global annual mean temperature curve started to increase, that the global cooling scenario was questioned. In the late 1980s the global annual mean temperature curve rose so steeply that all the dormant evidence from the late 1950s and 1960s was given prominence and the global warming theory was in full swing. What is intriguing is that some of the most vocal advocates for the global warming theory were also the ones responsible for creating

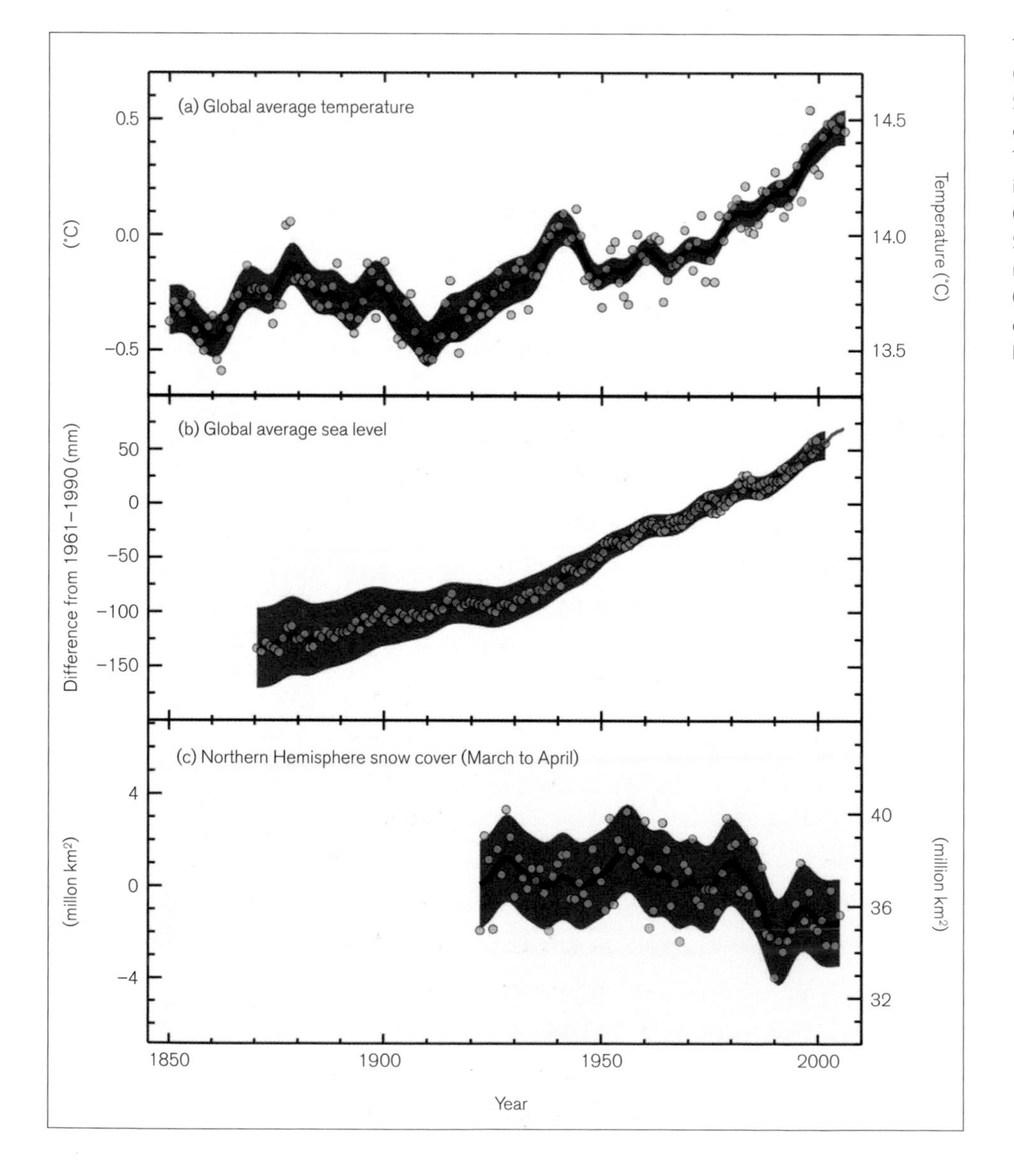

The Intergovernmental Panel on Climate Change (IPCC) report in 2007 compiled all the climate data covering the last 150 years. The data shows a temperature increase of over 0.75°C (1.3°F) (a), sea level rise of at least 22 cm (8.6 inches) (b) and a reduction of 3 million sq. km (1 million sq. miles) of snow cover in the Northern Hemisphere.

concern over the impending glacial period. In *The Genesis Strategy* in 1976, Stephen Schneider stressed that the global cooling trend had set in; he is now one of the leading proponents of global warming. In 1990 he stated that 'the rate of change [warming] is so fast that I don't hesitate to call that kind of change potentially catastrophic for ecosystems'.

This transition is very neatly described in John Gribbin's 1989 book *Hothouse Earth: The Greenhouse Effect and Gaia.*

> In 1981 it was possible to stand back and take a leisurely look at the record from 1880 to 1980 ... In 1987, the figures were updated to 1985, chiefly for neatness of adding another half a decade to the records ... But by early 1988, even one more year's worth of data justified another publication in April, just four months after the last 1987 measurements were made, pointing out the record-breaking warmth now being reached. Even there, Hansen [James Hansen, head of the NASA team studying global temperature trends] and Lebedeff were cautious about making the connection with the greenhouse effect, merely saying that this was 'a subject beyond the scope of this paper'. But in the four months it had taken to get the 1987 data in print, the world had changed again; just a few weeks later Hansen was telling the US Senate that the first five months of 1988 had been warmer than any comparable period since 1880, and the greenhouse effect was upon us.

It seems, therefore, that the whole global warming issue was driven by the upturn in the global annual mean temperature data set. We now know that the cooling trend of the 1960s and 1970s was due to the influence of the sunspot cycle (the intensity of sunspots varies over an 11-year period) and pollutants, such as sulphur dioxide aerosols, cooling the industrial regions of the globe. The upturn in the global annual mean temperature data, however, was not the sole reason for the appearance of the global warming issue, because in the late 1970s and 1980s there were significant advances in global climate modelling and a significant improvement in our understanding of past climates. The improvement in General Circulation Models (GCMs) during this period included taking into account the role of particles, clouds and ocean exchange of carbon dioxide in affecting the global climate. Despite the cooling effect thought to be associated with particle pollution, the new ocean–atmosphere coupled GCM tools emerged with revised and higher estimates of the warming that would be associated with a doubling of carbon dioxide in the atmosphere. By the 1980s, scientific concern had arisen about methane and other non-carbon dioxide greenhouse gases as well as the role of the oceans as a carrier of heat. GCMs continued to improve and the numbers of scientific teams working on such models increased over the 1980s and 1990s. In 1992, a first overall comparison of results from 14 GCM models was undertaken; the results were all in rough agreement, confirming the prediction of global warming.

When will the next glaciation happen?

During the 1980s there was also an ever-increasing drive to understand how and why past climate changed. Major advances were made in obtaining high-resolution past climate records from deep-sea sediments and ice cores. It was thus realized that glacial periods take tens of thousands of years to occur, primarily because ice sheets are very slow to build up and are naturally unstable. In contrast, the transition to a warmer period or interglacial, such as the present, is geologically very quick, in the order of a few thousand years. This is because once the ice sheets start to melt there are a number of positive feedbacks that accelerate the process, such as sea-level rise, which can undercut and destroy large ice sheets very quickly. The realization occurred in the palaeoclimate community that global warming is much easier and more rapid than cooling. It also put to rest the myth of the impending next glaciation. As the glacial–interglacial periods of the last 2.5 million years have been shown to be forced by the changes in the orbit of the earth around the sun, it would be possible to predict when the next glacial period will begin, if there were no human effects involved. According to the model predictions by Professor André Berger and his team at the Université catholique de Louvain in Belgium, we do not need to worry about another glacial period for at least 5,000 years. Indeed, if their model is correct and atmospheric carbon dioxide concentrations double, then global warming would postpone the next glacial for another 45,000 years. Professor Bill Ruddiman argues that the observed rise in atmospheric methane and carbon dioxide in the last 5,000 years

One of the biggest climate change concerns is the rate of melting of the ice sheets on Greenland and Antarctica. Scientists have found that both the Greenland and the Western Antarctic ice sheets (pictured below) are melting at a faster rate than predicted by the computer models.

could be due to early deforestation and agriculture. This, he argues, may have already stopped the modest global climate cooling that occurs before the climate steps down into full glacial conditions. Palaeoclimate work has also provided us with worrying insights into the speed at which climate systems can work. Recent studies of ice cores and deep-sea sediments demonstrate that at least regional climate changes of 5°C (9°F) can occur in a matter of decades. This work on reconstructing past climate seems to demonstrate that the global climate system is not benign but highly dynamic and prone to rapid changes.

Lessons from the past

Climate change in the geological past provides the context for the current concern regarding human-induced global warming, since by studying the past we can understand the mechanisms of climate change and its speed. Various geological records reveal that over the last 100 million years the earth's climate has been cooling down, moving from the so-called 'greenhouse world' of the Cretaceous period, when dinosaurs enjoyed warm and gentle conditions, through to the cooler and more dynamic 'icehouse world' of today. This long-term, 100-million-year transition to colder global climate conditions was driven mainly by tectonic changes, i.e. movement of the continents. These, as we saw in Chapter 3, included opening of the Tasmanian–Antarctic ocean gateway and the Drake Passage, which isolated Antarctica from the rest of the world; the uplift of the Himalayas; and the closure of the Panama ocean gateway. There is also geological evidence that this cooling has been accompanied by a massive drop in the levels of atmospheric carbon dioxide. One hundred million years ago, during the time of the dinosaurs, atmospheric carbon dioxide levels could in fact have been as much as five times higher than today.

In Chapter 3 we saw that this long-term cooling culminated in the glaciation of Antarctica about 35 million years ago and then the great Ice Age, which began 2.5 million years ago. At the beginning of this Ice Age, these glacial–interglacial cycles occurred every 41,000 years and since 1 million years ago they have occurred every 100,000 years. Our present interglacial, the Holocene period (see Chapter 7), started about 10,000 years ago and is an example of the warm conditions that occur between each glacial period. The Holocene began with the rapid and dramatic end of the last glacial; in less than 4,000 years global temperatures increased by 6°C (11°F), relative sea level rose by 120 m (394 ft), atmospheric carbon dioxide increased by a third and atmospheric methane doubled.

It may seem strange, considering the current obsession with global warming, to suggest that we are currently in a geological 'icehouse world'. This is because, despite

being in a relatively warm interglacial period, both poles are still glaciated, which is a rare occurrence in the geological history of our planet. Antarctica and Greenland are covered by ice sheets, and the majority of the Arctic Ocean is covered with sea ice. This means that there is a lot of ice that could melt in a warmer world, and, as we will see, this is one of the biggest unknowns that the future holds for our planet. The two glaciated poles also make the temperature gradient or difference between the poles and the Equator extremely large, from an average of about +30°C (86°F) at the Equator down to –35°C (–31°F) or colder at the poles. This temperature gradient is one of the main reasons that we have a climate system, as excess heat from the tropics is exported via both the oceans and the atmosphere to the poles. Geologically, we currently have one of the largest Equator–pole temperature gradients, which leads to a very dynamic climate system. So our 'icehouse' conditions cause our very energetic weather system, which is characterized by hurricanes, tornadoes, extra-tropical (temperate) winter storms and intense monsoons. In his book *The Ages of Gaia*, James Lovelock suggests that interglacials, such as the present Holocene period, are the 'fevered state' of our planet – over the last 2.5 million years the earth has shown that it actually prefers a colder average global temperature. Lovelock sees global warming as humanity just adding to the fever.

opposite A NASA satellite photograph of Hurricane Katrina, which hit New Orleans in 2005. Though it is impossible to attribute individual storms to global warming, there is a worrying trend towards more and bigger hurricanes in the Atlantic Ocean.

below The devastation left in New Orleans by Hurricane Katrina. The damage was done not by the hurricane winds but by the huge amount of rainfall which burst the levees constructed to protect the city, flooding vast areas.

The Little Ice Age

Climate, however, has not been constant during our present interglacial, i.e. the last 10,000 years. Palaeoclimatic evidence suggests that the early Holocene was warmer than the 20th century. Also, as we saw in Chapter 7, throughout the Holocene there have been millennial-scale climate events, called Bond events (the equivalent of the Dansgaard-Oeschger cycles discussed in Chapter 4), which involve a local cooling of 2°C (4°F), and can have a serious impact on human societies. The last of these climate cycles was the Little Ice Age. This was really two cold periods; the first followed the Medieval Warm Period which ended a thousand years ago, and is often referred to as the Medieval Cold Period. The Medieval Cold Period played a role in extinguishing Norse colonies on Greenland and caused famine and mass migration in Europe. It started gradually before AD 1200 and ended at about AD 1650. The second cold period, occurring between AD 1300 and 1860, is more classically referred to as the Little Ice Age, and may have been the greatest and most rapid change in the North Atlantic region during the late Holocene, as suggested by ice-core and deep-sea sediment records. However, it is clear from records around the globe that the Little Ice Age and the Medieval Warm Period only occurred in northern Europe, northeast America and Greenland. So the Little Ice Age was a regional rather than a worldwide climate fluctuation and cannot be used as an argument against global warming. It is often said that global warming is just the world 'recovering' from the Little Ice Age, but this is wrong because most of the world never had a Little Ice Age, so there is nothing to 'recover' from. The reconstructed global temperature records for the last millennium are essential as they provide a context for the instrumental temperature data set for the last 150 years. This clearly shows that temperatures, at least for the Northern Hemisphere, have been warmer in the 20th century than at any other time during the last 1,000 years, creating the so-called 'hockey stick' – with the past 1,000 years as the handle and the last 150 years as the blade.

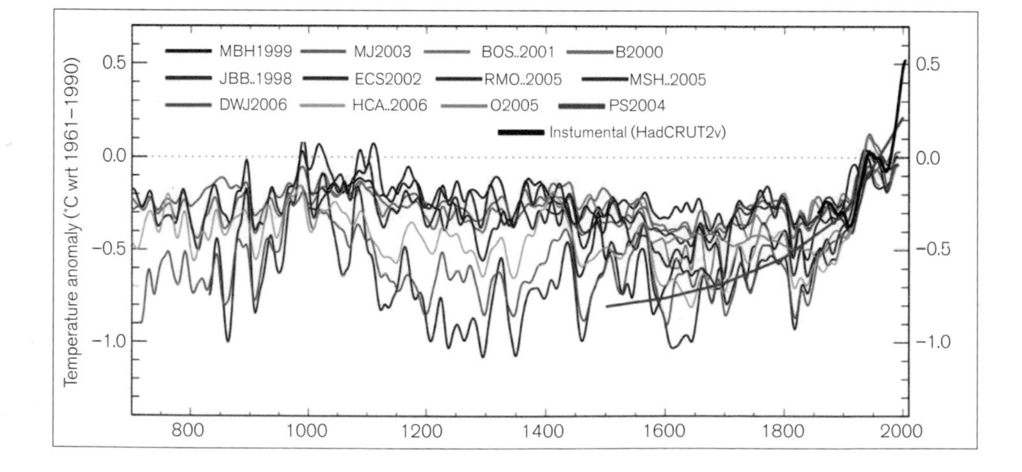

The IPCC report in 2007 compiled all the reconstructions of Northern Hemisphere temperature over the past 1,200 years. The resulting graph shows the sharp rise in temperatures during the 20th and 21st centuries and the shape of this graph is now referred to as the 'climate change hockey stick'.

The greenhouse effect

The temperature of the earth is determined by the balance between energy input from the sun and its loss back into space. Of the earth's incoming solar short-wave radiation (ultraviolet radiation and that in the visible spectrum), about a third is reflected back into space. The remainder is absorbed by the land and oceans, which radiate their acquired warmth as long-wave 'infrared' radiation. Atmospheric gases such as water vapour, carbon dioxide, ozone, methane and nitrous oxide are known as greenhouse gases and can absorb some of this long-wave radiation, becoming warmed by it. We need this greenhouse effect: without it, the earth would be at least 35°C (63°F) colder.

Plants take up water and carbon dioxide and, via photosynthesis, use solar energy to create the molecules they need for growth. Some of these plants are eaten by animals. Whenever the plants or animals die, they decompose and the retained carbon is released back into the 'carbon cycle', most returning into the atmosphere in gaseous form. However, if organisms die and are not allowed to rot, the embedded carbon is retained. Over a period of some 350 million years (but mainly in the so-called Carboniferous period, 359 to 299 million years ago), plants and small marine organisms died and were buried and crushed beneath sediments, forming 'fossil fuels' such as oil, coal and natural gas. The Industrial Revolution saw the start of large-scale combustion of these fossil fuels, releasing the carbon back into the atmosphere and as a result the temperature of the earth is rising.

Eighty per cent of all human-induced greenhouse gases entering the atmosphere come from industry, including the burning of coal and natural gas to produce electricity.

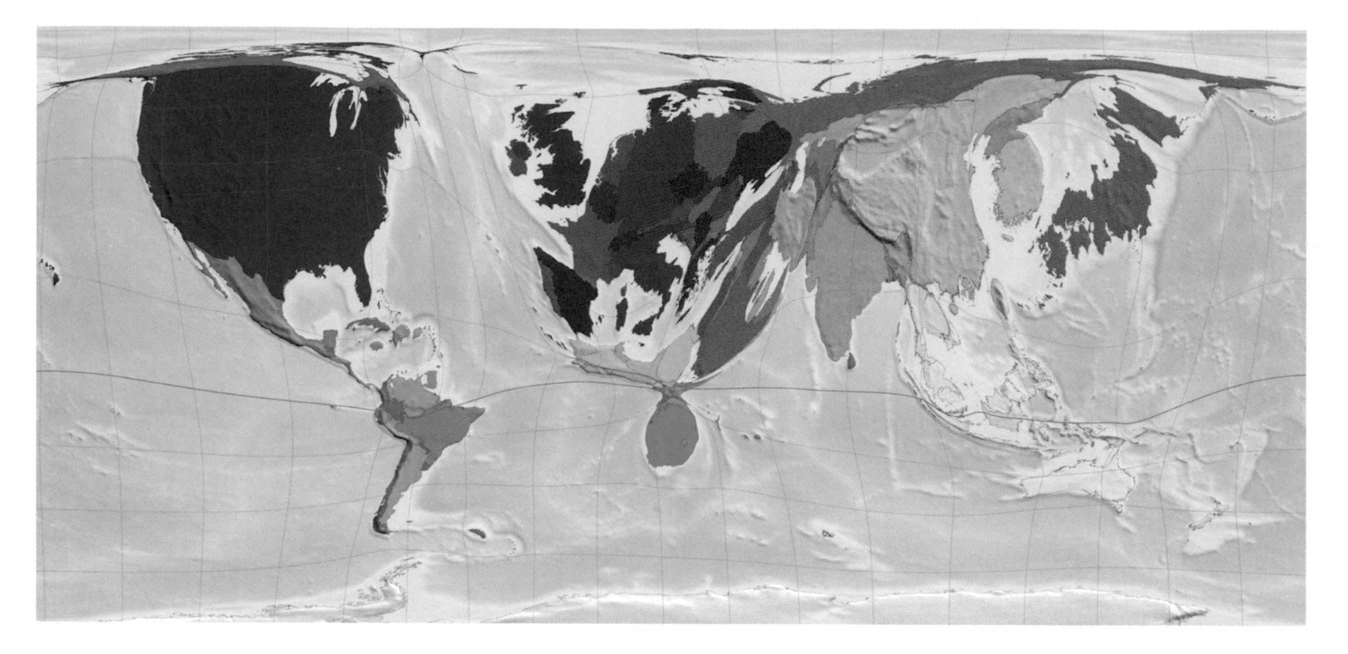

Past climate and the role of carbon dioxide

One of the ways in which we know that atmospheric carbon dioxide is important in controlling global climate is through the study of our past climate, particularly the glacial periods. The waxing and waning of the ice sheets is extremely rapid compared to other geological variations, such as the movement of continents around the globe, which occur over a time period of millions of years. But how do

above The production of greenhouse gases is not evenly distributed around the world. This digitally generated map – or cartogram – depicts the countries of the world not by their actual size but by the size of their carbon dioxide emissions in 2000. The cartogram clearly shows that it is the rich, developed countries of the world that produce the most carbon dioxide pollution.

left Climate scientists take samples of ice to measure the amount of pollution. Taking deep ice cores is how we have found that the natural level of carbon dioxide in the atmosphere during an interglacial is about 280 ppm. We have already increased this level to 387 ppm in less than 150 years.

we know about the role of carbon dioxide in these glaciations? The evidence, as we saw in Chapter 2, mainly comes from ice cores drilled in Antarctica and Greenland. As snow falls, it is light and fluffy and contains a lot of air. When this is slowly compacted to form ice, some of this air is trapped. By extracting these air bubbles trapped in the ancient ice, scientists can measure the percentage of greenhouse gases that were present in the past atmosphere. Scientists have drilled over 3 km (2 miles) down into both the Greenland and Antarctic ice sheets, which has enabled them to reconstruct the amount of greenhouse gases that occurred in the atmosphere over the last 0.5 million years. By examining the oxygen and hydrogen isotopes in the ice core, it is possible to estimate the temperature at which the ice was formed. The results are striking, as greenhouse gases such as atmospheric carbon dioxide and methane co-vary with temperatures over the last 650,000 years. This strongly supports the idea that the carbon dioxide content in the atmosphere and global temperature are closely linked, i.e. when carbon dioxide and methane increase, the temperature is found to increase and vice versa.

Human-induced climate change

The first direct measurements of atmospheric carbon dioxide concentrations started in 1958 at an altitude of about 4,000 m (13,000 ft) on the summit of Mauna Loa in Hawaii, a remote site free from local pollution. To extend this record further back, air bubbles trapped in ice are analysed. These long ice-core records suggest that pre-industrial carbon dioxide concentrations were about 280 parts per million (ppm). In 1958 the concentration was already 316 ppm, and it has climbed each and every year to reach 387 ppm by 2008. This level of pollution we have caused in one century is thus comparable to the natural waxing and waning of the glacial periods which, as we have seen, took thousands of years.

According to the Intergovernmental Panel on Climate Change (IPCC) report in 2007, increase in greenhouse gases over the last 150 years has already significantly changed the climate: average global temperatures have risen 0.75°C (1.3°F) and we have seen sea-level rises of over 22 cm (8.6 inches), significant shifts in the seasonality and intensities of precipitation, changing weather patterns and significant retreat of Arctic sea ice and nearly all continental glaciers. We know from records of the last 150 years that the 12 warmest years on record have all occurred in the last 13 years: 1998 was the warmest, followed by 2005, 2002, 2003 and 2004. The 8th warmest year was 2007. In 2007 the IPCC stated that the evidence for global warming is unequivocal and there is very high confidence that this is due to human activity. This view is supported by a vast array of scientific organizations, including the Royal Society and American Association for the Advancement of Science.

overleaf This aerial photograph shows the melting of Himalayan glaciers in Bhutan (white areas are where snow and ice still remain). The loss of ice in the Himalayas will soon severely affect water supplies in Pakistan and India and all the way round to China, potentially increasing regional tensions.

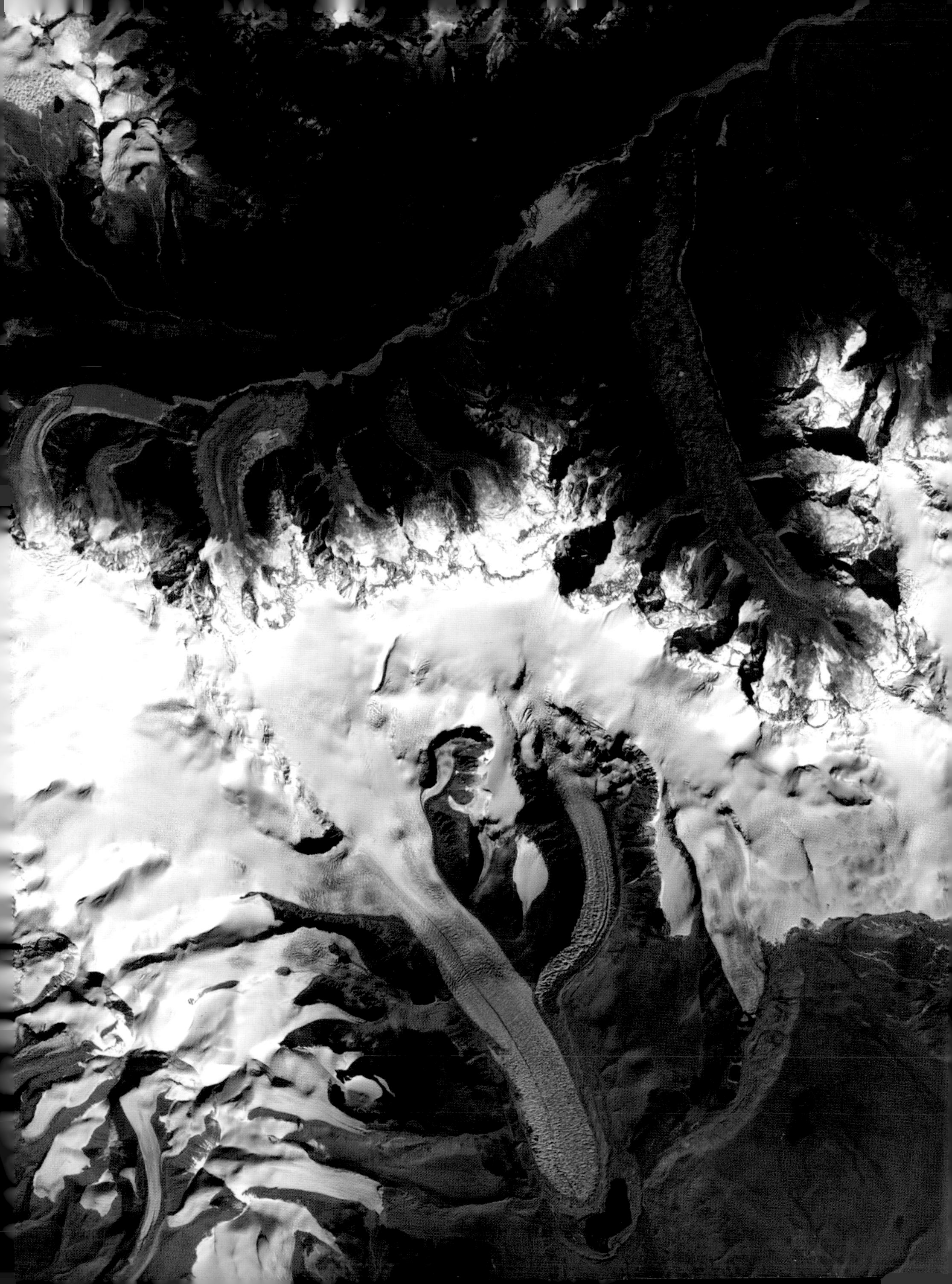

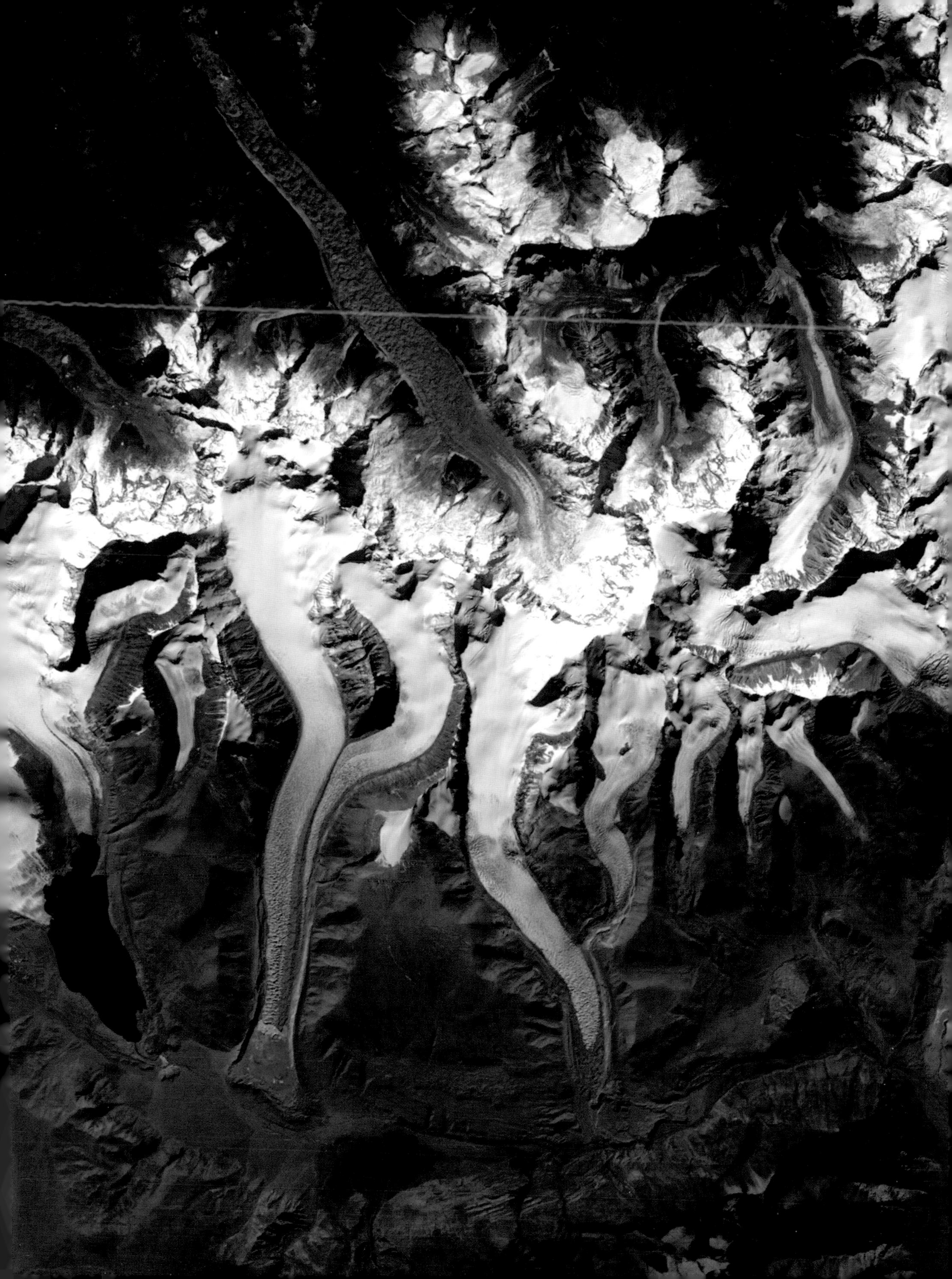

How do you model the future?

It may sound strange, but the whole of human society operates on knowing the future, and in particular the weather. For example, a farmer in India knows when the monsoon rains will come next year and so when to plant his crops, while a farmer in Indonesia knows there are two monsoon rains next year so he can plant crops twice. This is based on their knowledge of the past, as the monsoons have always come at about the same time each year in living memory. But such predictions go deeper than this, and influence every part of our lives. Our houses are built for the local climate – in England that means central heating but no air conditioning, while in the southern USA it is vice versa. Roads, railways, airports, offices, cars and trains are all designed for the local climate. This is why in the spring of 2003 a centimetre (half an inch) of snow one afternoon effectively shut down London, while Toronto can easily deal with and function with half a metre (1.6 ft) of snow. In Europe in 2003, 35,000 people died in the summer heat wave in temperatures that regularly occur in the tropics, while Australians go into shock if the temperature drops below 0°C (32°F). The problem with global warming is that it changes the rules. The past weather of an area cannot be relied on to tell you what the future will hold. We have to develop new ways of predicting the future, so that we can plan our lives and so that human society can continue to function fully. We have to model the future.

Climatologists use a whole hierarchy of climate models, from relatively simple models where different parts of the climate system are represented just by one to two

The UK Met Office Hadley Centre climate model has been used to predict temperature increases by the year 2100. Below you can see that the large temperature rises will have a disproportional distribution, with the greatest increases (shown as red areas) at the high latitudes.

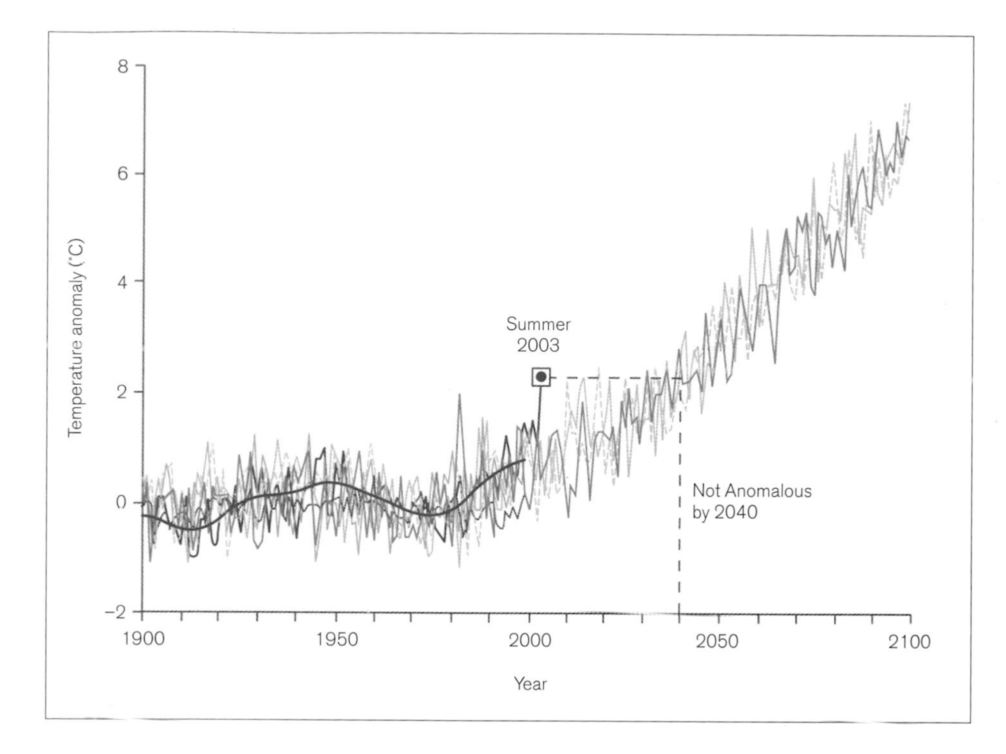

This graph depicts a comparison of model predictions of northern European summer temperatures and the real data. It shows the 2003 heat wave was anomalous, but owing to climate change average European summer temperatures will be at this level in less than 40 years.

boxes to the extremely complex three-dimensional General Circulation Models (GCMs). Each has a role in examining and furthering our understanding of the climate system. However, it is the GCMs that are used to predict future global climate. These comprehensive climate models are based on physical laws represented by mathematical equations that are solved using a three-dimensional grid over the globe. To obtain the most realistic simulations, all the major parts of the climate system must be represented in sub-models, including atmosphere, ocean, land surface (topography), cryosphere and biosphere, as well as the processes that go on within them and between them. As well as the warming effects of the greenhouse gases, the earth's climate system is complicated in that there are also cooling effects, such as the ice-albedo feedback (see Chapter 4). One of the main legacies of the Ice Age is the large amount of ice on the planet. The ice reflects most of the sunlight that falls on to it back into space, and this reflectiveness of the ground is called albedo. Predicting what will happen to the extent of global ice in the future creates huge difficulties in calculating the exact effect of global warming. For example, if the polar ice cap melts, the albedo will be significantly reduced, as this ice would be replaced by vegetation or open water, both of which absorb heat rather than reflecting it like white snow or ice. This would produce a positive feedback, enhancing the effects of global warming. There are already warning signs that the legacy of the Ice Age is starting to disappear. In the summer of 2007 the sea ice in the Arctic Ocean retreated further north than ever previously recorded.

What is the IPCC?

The Intergovernmental Panel on Climate Change (IPCC) was established in 1988 jointly by the United Nations Environmental Panel and the World Meteorological Organization because of worries about the possibility of global warming. The purpose of the IPCC is the continued assessment of the state of knowledge on the various aspects of climate change, including scientific, environmental and socio-economic impacts and response strategies. The IPCC does not undertake independent scientific research, it brings together all key research published in the world, synthesizes it and produces a consensus.

The IPCC is therefore recognized as the most authoritative scientific and technical voice on climate change, and its assessments have had a profound influence on the negotiators of the United Nations Framework Convention on Climate Change (UNFCCC) and its Kyoto Protocol. The meetings in The Hague in November 2000 and in Bonn in July 2001 were the second and third attempts to ratify (i.e. to make legal) the Protocols laid out in Kyoto in 1998. Unfortunately, President Bush pulled the USA out of the negotiations in March 2001. However, 191 other countries recognized by the UN made history in July 2001 by agreeing the most far-reaching and comprehensive environmental treaty the world has ever seen. The Kyoto Protocol entered into force finally on 16 February 2005, as it could only come into effect when Russia ratified the treaty, thereby meeting the requirement that at least 55 countries, representing 55 per cent of the global emissions, signed up to it. In December 2007 the newly elected Labour Prime Minister Kevin Rudd of Australia signed the Kyoto Protocol which was met with a standing ovation at the Bali meeting. As of April 2008, 178 countries out of a total of 192 recognized by the UN had ratified the treaty, leaving the USA as the only major country not to have signed up to Kyoto.

The IPCC is organized into three working groups plus a task force to calculate the amount of greenhouse gases produced by each country. Each of these four bodies has two co-chairmen (one from a developed and one from a developing country) and a technical support unit. Working Group I assesses the scientific aspects of the climate system and climate change; Working Group II addresses the vulnerability of human and natural systems to climate change, the negative and positive consequences of climate change, and options for adapting to them; and Working Group III assesses options for limiting greenhouse gas emissions and otherwise mitigating climate change as well as economic issues. Hence the IPCC also provides governments with scientific, technical and socio-economic information relevant to evaluating the risks and to developing a response to global climate change. The latest reports from these three working groups were published in 2007, and approximately 400 experts from some 120 countries were directly involved in drafting, revising and finalizing the IPCC reports. Another 2,500 experts participated in the review process. The IPCC authors are always nominated by governments and by international organizations including non-governmental organizations. These reports are essential reading for anyone interested in global warming (see Further Reading). In 2008 the IPCC was jointly awarded with Al Gore the Noble Peace Prize to acknowledge all the work the IPCC has done over the past 20 years.

In 2008 the Nobel Peace Prize was jointly awarded to the Intergovernmental Panel on Climate Change chaired by Professor Pachauri (right in picture) and Al Gore, to acknowledge their work on climate change.

Future climate change and its impact

The IPCC Report in 2007 synthesized the results of 23 Atmosphere–Ocean General Circulation Models, described above, to predict future temperature rises, based on six emission scenarios. It reports that global mean surface temperature could rise by between 1.1 and 6.4°C (1.9 and 11.5°F) by 2100, with best estimates being 1.8°C to 4°C (3.2 to 7.2°F). However, it should be noted that global carbon dioxide emissions are already rising faster than the most dire of the IPCC emission scenarios. The models also predict an increase in global mean sea level of between 18 and 59 cm (7 and 23 inches). If the contribution from the melting of Greenland and Antarctica is included then this range increases to between 28 and 79 cm (11 and 31 inches) by 2100. All such predictions assume a linear response between global temperatures and ice sheet loss. This is unlikely, and sea-level rise could thus be much higher.

The impacts of global warming will increase significantly as the temperature of the planet rises. The return period and severity of floods, droughts, heat waves and storms will worsen. Coastal cities and towns will be especially vulnerable as sea-level

The heat wave of 2003 killed over 35,000 people in northern Europe, and was so extreme that the Rhine river dried up.

rise will increase the effects of floods and storm surges. The increase of extreme climate events coupled with reduced water security and food security will have a severe effect on public health, affecting billions of people. Global warming also threatens global biodiversity. Ecosystems are already being hugely degraded by habitat loss, pollution and hunting. The 2007 Millennium Ecosystem Assessment report suggested that three known species were becoming extinct each hour, whilst the 2008 Living Planet Index suggested that the global biodiversity of vertebrates had fallen by over a third in just 35 years, an extinction rate now 10,000 times faster than any observed in the fossil record. Global warming is likely to exacerbate such degradation. Economic impacts will be severe and mass migration and armed conflict may result.

What is a safe level of climate change?

So what level of climate change is 'safe'? In February 2005 the British government convened an international science meeting at Exeter, UK to discuss this very topic.

The 2004 Hollywood blockbuster film *The Day After Tomorrow* was seen by a huge viewing public of at least 21 million people in the USA alone. In the film, climate change happens in a matter of weeks rather than years. One of the most memorable scenes is the flooding of New York as the ice sheets melt – there is still enough ice on Antarctica and Greenland to raise global sea levels by 70 m (230 ft) if it melted. The good news is that this is extremely unlikely because the majority of this ice is locked safely within the very stable Eastern Antarctic ice sheet.

Their recommendation is that global warming must be limited to a maximum of 2°C (4°F) above the pre-industrial average temperature. Below this threshold it seems that there were both winners and losers due to regional climate change, but above this figure everyone seems to lose. However, it now seems likely that temperature rises will exceed this threshold: a rise of 0.75°C (1.3°F) has already occurred, and had we stopped all emissions in 2000 there would still be an associated 0.6°C (1.08°F) rise due to the inertia and feedbacks in the climate system. If the impacts of small temperature rises are profound, sustained global temperature rises of 5–6°C (9–11°F) would be horrific. They could lead to the loss of both the Greenland and Western Antarctic ice sheets by the middle of the next century, raising sea levels by up to 12 m (39 ft). The UK Environment Agency has plans to deal with a rise of 4.5 m (15 ft) through construction of a barrier across the mouth of the river Thames stretching 24 km (15 miles) from Essex to Kent. However, a rise of 13 m (42 ft) would mean the flooding and permanent abandonment of nearly all lower-lying coastal and riverside urban areas. At the moment a third of the world's population lives within 100 km (60 miles) of a shoreline and 13 of the world's 20 largest cities are located on a coast. Billions could be displaced in mass migration. The North Atlantic Ocean circulation could collapse, plunging Western Europe into a succession of severe winters followed by heat waves in summer. At least 3 billion people in the world would not have enough water. Billions more would face starvation. The risk of armed conflict would rise hugely. Public health systems around the world would collapse. Global biodiversity would be devastated.

What is the cost?

So what is the cost of saving the world? According to the UK Government-commissioned Stern Review on the Economics of Climate Change in 2006, if we do everything we can now and reduce global greenhouse gas emissions and ensure we adapt to the coming effects of climate change it will cost us only 1 per cent of world GDP every year. However, if we do nothing then the impacts of climate change could cost between 5 and 20 per cent of world GDP every year. These figures have been disputed. Some experts have argued the cost of converting the global economy to low carbon could cost more than 1 per cent of GDP because global emissions have risen faster than the worst predictions. In response Sir Nicolas Stern has recently revised his figure to 2 per cent of world GDP, while others argue that the costs could easily be offset by a global carbon-trading system. Others suggest that the impacts and the associated costs of global warming have been underestimated by IPCC and the Stern Review. Even if the cost-benefit of solving global warming is less than suggested by Stern, there is an undeniable ethical case of preventing the deaths of tens of millions of people and the increase in human misery for billions.

Solutions

Global warming is the major challenge for our global society. We must not underestimate the task ahead of us. The climate predictions of the IPCC in 2007 were based on carbon emission scenarios of the next 100 years which were realistic forecasts in 2000. We now know that the economic miracle in China will cause carbon emissions to rise between 11 and 13 per cent between 2000 and 2010 instead of the highest estimate used for Asia by the IPCC of about 3–5 per cent. In addition, the consensus approach used by the IPCC to get agreement of all parties means it is inherently conservative. This means that we should view the top estimates of climate change as the more likely scenario, so we are staring down the barrel of a gun with over 6°C (11°F) warming by 2100. The climate system is also not linear, so there will be major tipping points when significant climate changes occur very rapidly. The figure opposite collates the tipping points which climatologists believe are most likely to happen in the near future and will be most devastating. If we cannot reverse the current global emissions trends, all of these tipping points will occur in our future.

What are the solutions to global warming? First there must be an international political solution; without a post-2012 agreement we are looking at huge increases in global carbon emissions and devastating global warming. Any political agreement will have to include developing countries and protect their rapid economic growth, as it is a moral imperative that people in the poorest countries have the right to obtain the same lifestyle currently enjoyed by those in developed countries. We also need massive investment in alternative and renewable power sources and low-carbon technology, to provide the means of reducing world carbon emissions. We cannot pin all our hopes on global politics and clean energy technology, so we must prepare for the worst and adapt. If implemented now, a lot of the costs and damage that could be caused by changing climate can be mitigated. This requires nations and regions to plan for the next 50 years, something that most societies are unable to do because of the very short-term nature of politics.

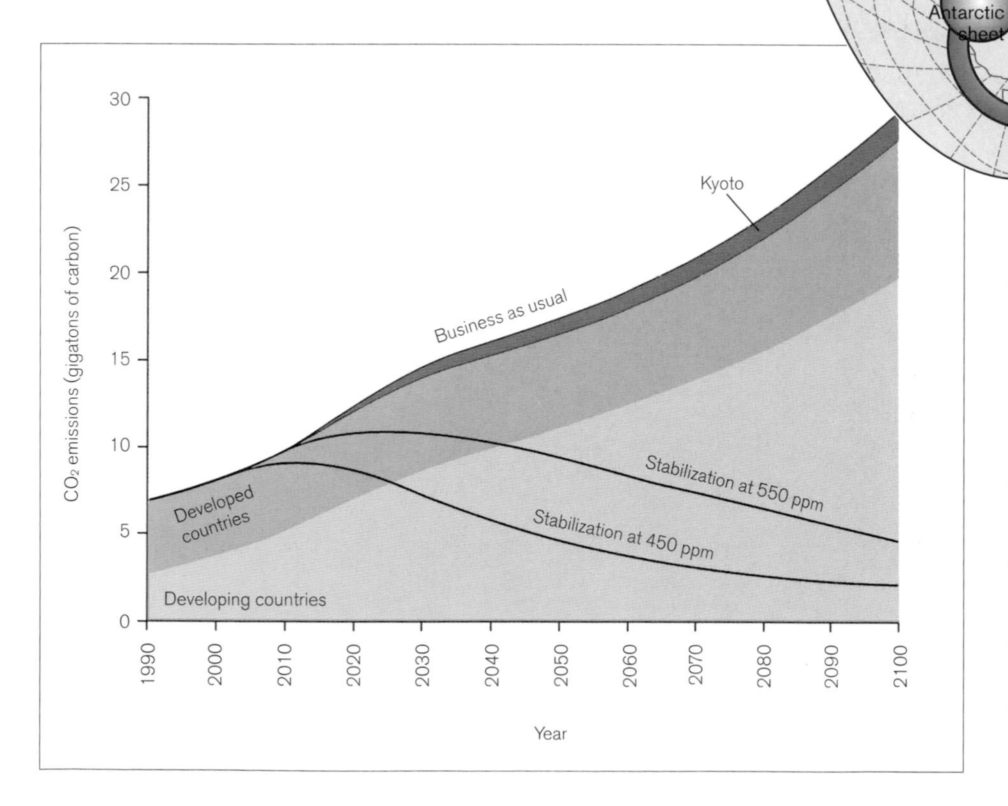

above A map of the potential climate change tipping points that could happen in the next 100 years.

left The predicted carbon dioxide emissions for a 'business-as-usual' world (in which current emission trends continue) and stabilization at atmospheric concentrations of either 550 ppm or 450 ppm. Many scientists believe that if we are to keep climate change below 2°C (4°F) we must aim for 450 ppm, but we are already at 387 ppm and emissions could rise in the future by over 2 ppm per year.

So global warming throws into question the very way we organize our society. Not only does it challenge the concept of the nation-state versus global responsibility, but also the short-term vision of our political leaders. To answer the question of what we can do about global warming, we must change some of the basic rules of our society to allow us to adopt a much more global and long-term sustainable approach.

Conclusions

The Ice Age has been the defining climate of the last 2.5 million years, and its legacy is all around us. The huge ice sheets of Greenland and Antarctica are testament to the coldness of our current global climate. Ironically, it is this ice that makes our climate uniquely sensitive to the increase in greenhouse gases. Humanity's greatest scientific experiment – what happens if you put lots of greenhouse gases into the atmosphere of a planet with lots of ice sheets – is under way. We should not despair as there are technological and political solutions to slow and even stop this great experiment, but we have to choose to use them. What the study of the Ice Age does tell us though, is that when climate changes it can do so suddenly and without warning.

opposite Alternative energy sources are essential for the future of the planet, as they will reduce carbon dioxide emissions. Wind turbines are an efficient means of generating electricity, if they are big enough. One study suggests that wind could, in principle, generate over 125,000 terrawatt-hours, which is five times the current global electricity requirement.

below Solar panels generate electricity when individual rays of the sun hit the panel and dislodge electrons inside it. The main advantage of solar panels is that you can place them where the energy is needed and avoid all the complicated infrastructure normally required to move electricity around.

Further Reading

There is a vast scholarly literature on ancient climate change and geology, the Ice Age bestiary and global warming. Here we indicate those titles that the non-specialist might wish to turn to first as supplementary reading on the various topics and themes discussed in this volume.

1 Discovering the Ice Age

Bahn, P. (ed.) *The Cambridge Illustrated History of Archaeology*, Cambridge University Press, Cambridge 1996

Imbrie, J. and Palmer Imbrie, K. *Ice Ages: Solving the Mystery*, Harvard University Press, Cambridge, MA 1979

Lurie, E. *Louis Agassiz: A Life in Science*, Johns Hopkins University Press, Baltimore 1988

2 Searching for Clues

Berger, A.J. et al. (eds) *Milankovitch and Climate: Understanding the Response to Astronomical Forcing*, D. Reidel, Dordrecht 2007

Flint, R.F. *Glacial and Quaternary Geology*, John Wiley, New York 1971

Imbrie, J. and Palmer Imbrie, K. *Ice Ages: Solving the Mystery*, Harvard University Press, Cambridge, MA 1979

Ruddiman, W. *Earth's Climate: Past and Future*, 2nd edition, W.H. Freeman, New York 2007

3 How the Age of Ice Began

Alley, R.B. *The Two-Mile Time Machine: Ice Cores, Abrupt Climate Change and our Future*, Princeton University Press, Princeton, NJ 2002

Christopherson, R.W. *Geosystems: An Introduction to Physical Geography*, 7th Edition, Prentice Hall, Upper Saddle River, NJ 2005

Corfield, R. *Architects of Eternity: The New Science of Fossils*, Headline Publishing, London 2001

Ruddiman, W.F. *Earth's Climate: Past and Future*, 2nd edition, W.H. Freeman, New York 2007

Seidov, D., Haupt, B.J. and Maslin, M.A. (eds) *The Oceans and Rapid Climate Change: Past, Present and Future*, AGU Geophysical Monograph Series Volume 126, 2001

Williams, M. et al., *Quaternary Environments*, 2nd edition, Edward Arnold, London 1998

4 The Climatic Rollercoaster

Alley, R.B. *The Two-Mile Time Machine: Ice Cores, Abrupt Climate Change and our Future*, Princeton University Press, Princeton, NJ 2002

Anderson, D.E., Goudie, A.S. and Parker, A.G. *Global Environments Through the Quaternary: Exploring Environmental Change*, Oxford University Press, Oxford and New York 2007

Lowe, J. and Walker, M. *Reconstructing Quaternary Environments*, 2nd edition, Prentice Hall, NJ 1997

Maslin, M.A. 'Quaternary Climate Thresholds and Cycles', *Encyclopedia of Paleoclimatology and Ancient Environments*, Kluwer Academic Publishers Earth Science Series 841–855, 2008

Maslin, M.A., Mahli, Y., Phillips, O. and Cowling S. 'New Views on an Old Forest: Assessing the Longevity, Resilience and Future of the Amazon Rainforest.' *Transactions of the Institute of British Geographers* 30, 4, 390–401, 2005

Ruddiman, W.F. *Earth's Climate: Past and Future*, 2nd edition, W.H. Freeman, New York, 2007

Williams, M. et al., *Quaternary Environments*, 2nd edition, Edward Arnold, London, 1998

Wilson, R.C.L., Drury S.A., and Chapman J.L., *The Great Ice Age: Climate Change and Life*, Routledge, London and New York 2003

5 The Human Story

Boaz, N.T. and Ciochon, R.L. *Dragon Bone Hill: An Ice-Age Saga of Homo erectus*, Oxford University Press, Oxford and New York 2004

Fagan, B.M. *The Great Journey: The Peopling of Ancient America*, Revised edition, University Press of Florida, Gainesville 2004

Fagan, B.M. *The Journey from Eden: The Peopling of Our World*, Thames & Hudson, London and New York 1990

Gamble, C. *Timewalkers: The Prehistory of Global Colonization*, Harvard University Press, Cambridge, MA 1994

Haynes, G. *Early Settlement of North America: The Clovis Era*, Cambridge University Press, Cambridge and New York 2002

Hoffecker, J.F. *A Prehistory of the North: Human Settlement of the Higher Latitudes*, Rutgers University Press, New Brunswick 2005

Hoffecker, J.F. and Elias, S.A. *Human Ecology of Beringia*, Columbia University Press, New York 2007

Lewin, R. *Human Evolution: An Illustrated Introduction*, Revised edition, John Wiley & Sons, New York 2004

Mellars. P. *The Neanderthal Legacy: An Archaeological Perspective from Western Europe*, Princeton University Press, Princeton, NJ 1996

Mithen, S. *The Prehistory of the Mind: The Cognitive Origins of Art and Science*, Thames & Hudson, London and New York 1996

Pitts, M. and Roberts, M. *Fairweather Eden: Life in Britain Half a Million Years Ago as Revealed by the Excavations at Boxgrove*, Random House UK, London 1998

Stringer, C. and Andrews, P. *The Complete*

World of Human Evolution, Thames & Hudson, London and New York 2005
Stringer, C. *Homo Britannicus*, Allen Lane, London 2006
Stringer, C. and Gamble, C. *In Search of the Neanderthals: Solving the Puzzle of Human Origins*, Thames & Hudson, London and New York 1993
Stringer, C. and McKie, R. *African Exodus: The Origins of Modern Humanity*, Henry Holt and Company, New York 1996
Swisher, C.C., Carl, C., Curtis, G.H. and Lewin, R. *Java Man: How Two Geologists' Dramatic Discoveries Changed Our Understanding of the Evolutionary Path to Modern Humans*, Scribner, New York 2000
Tattersall, I. *The Last Neanderthal: The Rise, Success, and Mysterious Extinction of Our Closest Human Relatives*, Westview Press, Boulder 1999
Tattersall, I. and Schwartz, J.H. *Extinct Humans*, Westview Press, Boulder 2000
Trinkaus, E. and Shipman, P. *The Neandertals: Of Skeletons, Scientists, and Scandal*, Vintage Books, New York 1992
Walker, A. and Shipman, P. *The Wisdom of the Bones: In Search of Human Origins*, Alfred A. Knopf, New York 1996
Wood, B. *Human Evolution: A Very Short Introduction*, Oxford University Press, Oxford and New York 2005

6 The Ice Age Bestiary

Flannery, T. *The Eternal Frontier: An Ecological History of North America and its Peoples*, Penguin, London 2001 and Vintage, New York 2002
Guthrie, R.D. *Frozen Fauna of the Mammoth Steppe: The Story of Blue Babe*, Columbia University Press, New York 1990
Lange, I.M. *Ice Age Mammals of North America: A Guide to the Big, the Hairy and the Bizarre*, Mountain Press Publishing Company, Missoula, MT 2002.
Lister, A. and Bahn, P. *Mammoths: Giants of the Ice Age*, Frances Lincoln, London 2007
Long, J., Archer, M., Flannery, T. and Hand, S. *Prehistoric Mammals of Australia and New Guinea: One Hundred Million Years of Evolution*, Johns Hopkins University Press, Baltimore 2002
Martin, P.S. *Twilight of the Mammoths: Ice Age Extinctions and the Rewilding of America*, California University Press, Berkeley 2005
Molnar, R.E. *Dragons in the Dust: The Paleobiology of the Giant Monitor Lizard Megalania*, Indiana University Press, Bloomington and Indianapolis 2004
Turner, A. and Antón, M. *The Big Cats and their Fossil Relatives*, Columbia University Press, New York 1997
Turner, A. and Antón, M. *Evolving Eden: An Illustrated Guide to the Evolution of the African Large-mammal Fauna*, Columbia University Press, New York 2004
Turner, A. and Antón, M. *Prehistoric Mammals*, National Geographic, Washington D.C. 2004

7 After the Ice

Bailey, G. and Spikins, P. (eds) *Mesolithic Europe*, Cambridge University Press, Cambridge 2008
Barker, G. *The Agricultural Revolution in Prehistory*, Oxford University Press, Oxford 2006
Diamond, J. *Collapse: How Societies Choose to Fail or Survive*, Penguin, London and New York 2006
Diamond, J. *Guns, Germs and Steel: A Short History of Everybody for the Last 13,000 Years*, Vintage, London 1998 and W.W. Norton, New York 1999
Fagan, B.M. *Floods, Famines and Emperors: El Niño and the Fate of Civilizations*, Basic Books, New York 2009
Fagan, B.M. *The Great Warming: Climate Change and the Rise and Fall of Civilizations*, Bloomsbury, London and New York 2008
Fagan, B.M. *The Long Summer: How Climate Changed Civilization*, Granta Books, London 2005, Basic Books, New York 2004
Fagan, B.M. *The Little Ice Age: How Climate Made History 1300–1850*, Basic Books, New York 2001
Mithen, S. *After the Ice: A Global Human History, 20,000–5,000 BC*, second edition, Orion, London 2004 and Harvard University Press, Cambridge, MA 2006
Renfrew, C. *Prehistory: The Making of the Human Mind*, Weidenfeld, London 2008 and Modern Library, New York 2008

8 Hot or Cold Future?

Corfee-Morlot, J., Maslin, M.A. and Burgess, J. 'Climate Science in the Public Sphere', *Philosophical Transactions A of the Royal Society*, 2007
Flannery, T. *The Weather Makers: Our Changing Climate and What it Means for Life on Earth*, Grove/Atlantic, New York 2006 and Penguin, London 2007
Gribbin, J. *Hothouse Earth: The Greenhouse Effect and Gaia*, Grove/Atlantic, New York, 1990
Houghton, J.T. *Global Warming: The Complete Briefing*, 3rd edition, Cambridge University Press, Cambridge 2004
Lovelock, J. *The Ages of Gaia: A Biography of Our Living Earth*, Oxford University Press, Oxford and New York 2000
Maslin, M. *A Very Short Introduction to Global Warming*, Oxford University Press, Oxford 2008
Metz et al. (ed.) IPCC *Climate Change 2007: Mitigation of Climate Change, Contribution of Working Group III to the Fourth Assessment Report of the Intergovernmental Panel on Climate Change*, Cambridge University Press, Cambridge 2007
Monbiot, G. *Heat*, Allen Lane, London 2006
Stern, N. *The Economics of Climate Change: The Stern Review*, Cambridge University Press, Cambridge 2007
Walker, G. and King, D. *The Hot Topic*, Bloomsbury, London 2008

Contributors

Brian M. Fagan is Emeritus Professor of Anthropology at the University of California, Santa Barbara. He has written or edited over 40 widely read books, many of them on ancient climate change, including *Floods, Famines and Emperors: El Niño and the Fate of Civilizations*; *The Little Ice Age: How Climate Made History; The Great Warming: Climate Change and the Rise and Fall of Civilizations* and *The Long Summer: How Climate Changed Civilization.* He is also the editor of *The Seventy Great Mysteries of the Ancient World*; *The Seventy Great Inventions of the Ancient World* and *Discovery! Unearthing the New Treasures of Archaeology*, all published by Thames & Hudson.

Mark Maslin is Professor of Geography at University College London and is the Director of the UCL Environment Institute. A leading palaeoclimatologist, he has particular expertise in past global and regional climatic change, as well as government climate change policies, causes of past and future global climate change, ocean circulation, Amazonia and East Africa. Maslin has published seven popular books, including the highly successful *Global Warming: A Very Short Introduction* (OUP), as well as over 90 academic papers. The first UCL Environment Institute Policy Report led by Professor Maslin was the basis of the Channel 4 'Dispatches' programme *Greenwash* in 2007.

John F. Hoffecker is a Fellow at the Institute of Arctic and Alpine Research at the University of Colorado. An expert on the evolution of human adaptations to cold environments, much of his research and writing has been focused on the Ice Age archaeology of Russia and Alaska, and in 2005 he received an honorary degree from the Russian Academy of Sciences. He is the author of numerous publications, including the books *Desolate Landscapes*; *A Prehistory of the North*, and with Scott A. Elias, *Human Ecology of Beringia.*

Hannah O'Regan is a Senior Research Officer at Liverpool John Moores University, UK. An archaeologist and palaeontologist, she has a particular interest in Ice Age carnivores and cave archaeology. She has published scientific papers on the dispersal of Quaternary mammals (including early humans and other primates), the fossil Carnivora of southern Africa, the history and archaeology of zoos, and cave archaeology in northern Britain. She has worked on a number of archaeological sites in the UK and South Africa, ranging in date from 2.5 million years ago to the early 20th century.

Sources of Illustrations

1 Stephen Morley; **2–3** The Natural History Museum, London; **4–5** Semitour Périgord; **7** Julia Razumovitch/ istockphoto.com; **8** © Roger Ressmeyer/Corbis; **1** Francesco Tomasinelli/Tips Images; **9** The Art Archive/Corbis; **10–11** Kenneth Garrett/National Geographic; **13** Benoit Audureau/The Natural History Museum, London;**14** Colin Monteath/Minden Pictures/National Geographic; **15** Jan Will/istockphoto.com; **16–17** Anthony Dodd/istockphoto.com; **18** Muséum de'histoire naturelle de Neuchâtel; **19** Charles Lyell, *A Second Visit to the United States*, 1849; **20** Goldenhawk/SnapVillage; **21** A. Geikie, *Life of Sir R I Murchison..with notices of his scientific contemporaries...*, 1875; **22–23** Louis Agassiz, *Etudes Sur Les Glaciers*, 1841; **25** Kenneth Garrett/ National Geographic; **26–27** The Natural History Museum, London; **28** Imagestate/ Tips Images; **29** James Geikie, *The Great Ice Age*, 1894 (3rd edition); **30–31** Eric Grave/Science Photo Library; **32** © Phil Schermeister/Corbis; **33** © Lowell Georgia/Corbis; **34** Simpson, *Eruption of Krakatoa*, 1888; **35** James Campbell Irons, *Autobiographical Sketch of James Croll*, 1875; **36** James Croll, *Climate and Time in their Geological Relations*, 1875; **37a** Vasko Milankovitch; **39a** Emilio Segre Visual Archives/American Institute of Physics/ Science Photo Library; **39b** James King-Holmes/Science Photo Library; **40** Des Bartlett/Science Photo Library; **41** © Dean Conger/Corbis **43** E. R. Degginger/ Science Photo Library; **44** Roman Krochuk/istockphoto.com; **45** Roger Harris/Science Photo Library; **47** IDOP; **48–49** Bill Grove/istockphoto.com; **51b** NASA Goddard Space Flight Center; **55** NASA; **56a** Science Photo Library; **56b** Mark Maslin; **60–61** NOAA/Science Photo Library; **57** TT/istockphoto.com; **62–63** Anne Jennings; **64** © Mike Zens/ Corbis; **65** David M. Anderson, NOAA Paleoclimatology Program and INSTARR, University of Colorado, Boulder; **67** ML Design **68** ML Design **70** ML Design **71** Viktor Glupov/istockphoto.com; **72–73** © Dominic Harcourt Webster/ Robert Harding World Imagery/Corbis; **74l** U.S. Geological Survey; **74r** Markus Divis/istockphoto.com; **75** ML Design **76–7** ML Design **79** Morley Read/ istockphoto.com; **78** Heikie Hofstaetter/ istockphoto.com; **85** ML Design **86** Nancy Nehring/istockphoto.com; **87** After John

T. Andrews and Thomas G. Andrews; **88** The Bedford Institute of Oceanography and Dalhousie University for the University of Colorado; **90** Anne Jennings; **91** ML Design **92** Javier Trueba/MSF/Science Photo Library; **92–93** Getty Images; **95** Konrad Wothe/Minden Pictures/National Geographic; **96** John Reader/Science Photo Library; **97** Javier Trueba/MSF/Science Photo Library; **98** Kenneth Garrett/National Geographic; **99** Kenneth Garrett/National Geographic; **100** Volker Steger/Nordstar – 4 Million Years of Man/Science Photo Library; **101** Mary Jelliffe/Ancient Art & Architecture; **102** John Sibbick; **103a** Photo RMN; **103b** John Sibbick; **104** © Boxgrove Project; **105a** ML Design, after Bocherens et al, 2005; **105b** Imagestate/Tips Images; **106a** © Boxgrove Project; **106r** Christa S. Fuchs, Niedersächisches Landesamt für Denkmalpflege; **107** © Boxgrove Project; **108–9** Philippe Plailly/Eurelios/Science Photo Library; **110** Courtesy Professor Naama Goren-Inbar; **111** John Sibbick; **112** ML Design **113** Kenneth Garrett/ National Geographic; **114** Kenneth Garrett/National Geographic; **115** Viktoria Kulish/istockphoto.com; **116** The Natural History Museum, London; **117** Jeremy Percival; **118l** Javier Trueba/MSF/Science Photo Library; **119** State Hermitage Museum, St Petersburg; **118r** Jay Maidment/The Natural History Museum, London; **120–121** ML Design **122** Courtesy Chris Henshilwood; **123** Courtesy Chris Henshilwood; **125a** Sisse Brimberg/ National Geographic; **125b** French Ministry of Culture and Communication, Regional Direction for Cultural Affairs – Rhône–Alpes region – Regional department of archaeology; **126** Thomas Stephan © Ulmer Museum; **127** State Hermitage Museum, St Petersburg; **128–9** John Sibbick; **130** State Hermitage Museum, St Petersburg; **131** akg-images/Erich Lessing; **132** Photo RMN; **133** Sisse Brimberg/National Geographic; **134–35** Semitour Périgord; **136** Kenneth Garrett/National Geographic; **137** Irina Igumnova/ istockphoto.com; **138–9** John Sibbick; **140** ML Design **141** Courtesy of Smithsonian Institution, Washington D.C.; **142–3** John Cancalosi/National Geographic; **145a** Arco Digital Images/Tips Images; **145b** The Natural History Museum, London; **146** © Gallo Images/Corbis; **147a** Ghar Dalam Museum, Malta; **147b** ML Design **149** John Reader/Science Photo Library; **150** Lee Pettet/istockphoto.com; **152** Mauricio Anton/Science Photo Library; **153** Philippe Plailly/ Eurelios/Science Photo Library; **154** French Ministry of Culture and Communication, Regional Direction for Cultural Affairs - Rhône-Alpes region - Regional department of archaeology; **155** Mauricio Anton/Science Photo Library; **156** Martin Raul/National Geographic; **157** Mauricio Anton/Science Photo Library; **158a** Richard Nowitz/National Geographic; **158b** Muséum Autun; **159** A.V. Lozhkin; **160** RIA Novosti; **161a** Michael Long/The Natural History Museum, London; **161b** ML Design **162** Jean Vertut; **163** Semitour Périgord; **164–5** French Ministry of Culture and Communication, Regional Direction for Cultural Affairs – Rhône–Alpes region – Regional department of archaeology; **166** The University of Leeds, School of Biology; **167** The Natural History Museum, London; **168** Martin Shields/Science Photo Library; **169** University of Alaska Museum; **171** The Natural History Museum, London; **172a** Mauricio Anton/Science Photo Library; **172b** Paleozoological Museum of China, Beijing; **173** Courtesy of Smithsonian Institution, Washington D.C.; **174** Martin Shields/Science Photo Library; **175** Tom McHugh/Science Photo Library; **176** W K Fletcher/Science Photo Library; **177** Jason Edwards/National Geographic; **178** South Australian Museum, Adelaide; **179a** Science Source/Science Photo Library; **179b** Tasmanian Archive and Heritage Office, State Library of Tasmania; **180** South Australian Museum; **181** South Australian Museum, Adelaide; **182** Richard Nowitz/National Geographic; **185** ML Design **186–7** © Paul Almasy/Corbis; **189** ML Design **190** Norbert Rosing/National Geographic; **191** The British Museum, London; **192** © Adam Woolfitt/Corbis; **193a** Marja Gimbutas; **193b** National Museum of Denmark, Copenhagen; **194** R. Workman; **195** © Hanan Isachar/Corbis; **196** Sisse Brimberg/National Geographic; **197a** Israel Antiquities Authority, Jerusalem; **197b** © Keren Su/Corbis; **198** Lowell Georgia/National Geographic; **199a** SnapVillage; **199b** The British Museum, London; **200** The British Museum, London; **201** © Georg Gerster/Panos Pictures; **202** Craig Chiasson/istockphoto.com; **203** Martin Gray/National Geographic; **205** The Art Archive/ Museum of London/Eileen Tweedy; **206–7** © Kennan Ward/Corbis; **208** Chromorange/Tips Images; **212** © Paul Souders/Corbis; **214**NASA; **215** © David J. Philip/epa/Corbis; **216** IPCC; **217** Michael Utech/istockphoto.com; **218a** From *The Atlas of the Real World: Mapping the Way We Live* by Daniel Dorling, Mark Newman and Anna Barford. © 2008 Daniel Dorling, Mark Newman and Anna Barford. Published by Thames & Hudson Ltd., London, 2008; **218b** Hidden Ocean 2005 Expedition, NOAA Office of Ocean Exploration; **220–1** Jeffrey Kargal USGS/NASA JPL/AGU; **222** NASA; **223** Mark Maslin; **224** © Bjorn Sigurdson/epa/Corbis; **225** © Martin Gerten/epa/Corbis; **226–7** TM 20th Century Fox/Album/AKG; **229** ML Design, after Mark Maslin; **230** drob/Snapvillage; **231** Elyrae/Snapvillage.

Index

Page numbers in *italic* refer to illustrations.